D0857658

Population Genetics

Population Genetics

A Concise Guide

John H. Gillespie

THE JOHNS HOPKINS UNIVERSITY PRESS
Baltimore and London

Published 1998
Printed in the United States of America on acid-free paper
07 06 05 04 03 02 01 00 99 98 5 4 3 2 1

The Johns Hopkins University Press
2715 North Charles Street
Baltimore, Maryland 21218-4319
The Johns Hopkins Press Ltd., London

Library of Congress Cataloging-in-Publication Data will be found at the end of this book.
A catalog record for this book is available from the British Library.

ISBN 0-8018-5754-6

ISBN 0-8018-5755-4 (pbk.)

To Robin Gordon

Contents

List of Figures

Preface

At various times I have taught population genetics in two- to five-week chunks. This is precious little time in which to teach a subject, like population genetics, that stands quite apart from the rest of biology in the way that it makes scientific progress. As there are no textbooks short enough for these chunks, I wrote a *Minimalist's Guide to Population Genetics.* In this 21-page guide I attempted to distill population genetics down to its essence. This guide was, for me, a central canon of the theoretical side of the field. The minimalist approach of the guide has been retained in this, its expanded incarnation. My goal has been to focus on that part of population genetics that is central and incontrovertible. I feel strongly that a student who understands well the core of population genetics is much better equipped to understand evolution than is one who understands less well each of a greater number of topics. If this book is mastered, then the rest of population genetics should be approachable.

Population genetics is concerned with the genetic basis of evolution. It differs from much of biology in that its important insights are theoretical rather than observational or experimental. It could hardly be otherwise. The objects of study are primarily the frequencies and fitnesses of genotypes in natural populations. Evolution is the change in the frequencies of genotypes through time, perhaps due to their differences in fitness. While genotype frequencies are easily measured, their change is not. The time scale of change of most naturally occurring genetic variants is very long, probably on the order of tens of thousands to millions of years. Changes this slow are impossible to observe directly. Fitness differences between genotypes, which may be responsible for some of the frequency changes, are so extraordinarily small, probably less than 0.01 percent, that they too are impossible to measure directly. Although we can observe the state of a population, there really is no way to explore directly the evolution of a population.

Rather, progress is made in population genetics by constructing mathematical models of evolution, studying their behavior, and then checking whether the states of populations are compatible with this behavior. Early in the history of population genetics, certain models exhibited dynamics that were of such obvious universal importance that the fact that they could not be directly verified in a natural setting seemed unimportant. There is no better example than genetic drift, the small random changes in genotype frequencies caused by variation in offspring number between individuals and, in diploids, genetic segregation. Genetic drift is known to operate on a time scale that is proportional to the size of the population. In a species with a million individuals, it takes roughly a million generations for genetic drift to change allele frequencies appreciably. There is no conceivable way of verifying that genetic drift changes allele frequencies in most natural populations. Our understanding that it does is entirely theoretical. Most population geneticists not only are comfortable with this state of affairs but also revel in the fact that they can demonstrate on the back of an envelope, rather than in the laboratory, how a significant evolutionary force operates.

As most of the important insights of population genetics came initially from theory, so too is this text driven by theory. Although many of the chapters begin with an observation that sets the biological context for what follows, the significant concepts first appear as ideas about how evolution ought to proceed when certain assumptions are met. Only after the theoretical ideas are in hand does the text focus on the application of the theory to an issue raised by experiments or observations.

The discussions of many of these issues are based on particular papers from the literature. I chose to use papers rather than my own summary of several papers to involve the reader as quickly as possible with the original literature. When I teach this material, I require that both graduate and undergraduate students actually read the papers. Although this book describes many of the papers in detail, a deep understanding can only come from a direct reading. Below is a list of the papers in the order that they appear in the text. I encourage instructors to make the papers available to their students.

1. CLAYTON, G. A., MORRIS, J. A., AND ROBERTSON, A. 1957. An experimental check on quantitative genetical theory. II. Short-term responses to selection. *J. Genetics* 55:131–151.

2. CLAYTON, G. A., AND ROBERTSON, A. 1955. Mutation and quantitative variation. *Amer. Natur.* 89:151–158.

3. GREENBERG, R., AND CROW, J. F. 1960. A comparison of the effect of lethal and detrimental chromosomes from *Drosophila* populations. *Genetics* 45:1153–1168.

4. HARRIS, H. 1966. Enzyme polymorphisms in man. *Proc. Roy. Soc. Ser. B* 164:298–310.

5. KIMURA, M., AND OHTA, T. 1971. Protein polymorphism as a phase of molecular evolution. *Nature* 229:467–469.

6. KIRKPATRICK, M., AND JENKINS, C. D. 1989. Genetic segregation and the maintenance of sexual reproduction. *Nature* 339:300–301.

7. KONDRASHOV, A. 1988. Deleterious mutations and the evolution of sexual reproduction. *Nature* 336:435–440.

8. KREITMAN, M. 1983. Nucleotide polymorphism at the alcohol dehydrogenase locus of *Drosophila melanogaster*. *Nature* 304:412–417.

9. Morton, N. E., Crow, J. F., and Muller, H. J. 1956. An estimate of the mutational damage in man from data on consanguineous marriages. *Proc. Natl. Acad. Sci. USA* 42:855–863.

Each chapter contains a short overview of what is to follow, but these overviews are sometimes incomprehensible until the chapter has been read and understood. The reader should return to the overview after mastering the chapter and enjoy the experience of understanding what was previously mysterious. Each chapter of the text builds on the previous ones. A few sections contain more advanced material, which is not used in the rest of the book and could be skipped on a first reading; these are sections 2.6, 2.7, 3.8, 5.4, and 5.5. Certain formulae are placed in boxes. These are those special formulae that play such a central role in population genetics that they almost define the way most of us think about evolution. Everyone reading this book should make the boxed equations part of their being.

Problems have been placed within the text at appropriate spots. Some are meant to illuminate or reinforce what came before. Others let the reader explore some new ideas. Answers to all but the most straightforward problems are given at the end of each chapter.

The prerequisites for this text include Mendelian genetics, a smattering of molecular genetics, a facility with simple algebra, and a firm grasp of elementary probability theory. The appendices contain most of what is needed in the way of mathematics, but there is no introduction to genetics. With so many good genetics texts available at all levels, it seemed silly to provide a cursory overview.

Many people have made significant contributions to this book. Among the students who suffered through earlier drafts I would like to single out Suzanne Pass, who gave me pages of very detailed comments that helped me find clearer ways of presenting some of the material and gave me some understanding of how the book sells to a bright undergraduate. Dave Cutler was my graduate teaching assistant for a 10-week undergraduate course based on an early draft. In addition to many invaluable comments, Dave also wrote superb answers to many of the problems. Other students who provided helpful comments included Joel Kniskern, Troy Thorup, Jessica Logan, Lynn Adler, Erik Nelson, and Caroline Christian. I regret that the names of a few others may have disappeared in the clutter on my desk. You have my thanks anyway.

Chuck Langley taught a five-week graduate course out of the penultimate draft. He not only found many errors and ambiguities but also made the genetics much more precise. Mel Green helped in the same way after a thorough reading from cover to cover (not bad for a man who looks on most of population genetics with skepticism!). Michael Turelli answered innumerable questions about quantitative genetics, including the one whose answer I hated: Is this how you would teach quantitative genetics? Monty Slatkin made many helpful suggestions based on a very early version. David Foote provided the data for Figure 5.1.

Finally, my greatest debt is to my wife, Robin Gordon, who not only encouraged me during the writing of this book but also edited the entire manuscript. More important, she has always been my model of what a teacher should be. Whatever success I may have had in teaching population genetics has been inspired in no small part by her. In keeping with the tradition established in my previous book of dedications to great teachers, I dedicate this one to her.

Population Genetics

Chapter 1

The Hardy-Weinberg Law

Population geneticists spend most of their time doing one of two things: describing the genetic structure of populations or theorizing on the evolutionary forces acting on populations. On a good day, these two activities mesh and true insights emerge. In this chapter, we will do all of the above. The first part of the chapter documents the nature of genetic variation at the molecular level, stressing the important point that the variation between individuals within a species is similar to that found between species. After a short terminologic digression, we begin the theory with the traditional starting point of population genetics, the Hardy-Weinberg law, which describes the consequences of random mating on allele and genotype frequencies. Finally, we see that the genotypes at a particular locus do fit the Hardy-Weinberg expectations and conclude that the population mates randomly.

No one knows the genetic structure of any species. Such knowledge would require a complete description of the genome and spatial location of every individual at one instant in time. In the next instant, the description would change as new individuals are born, others die, and most move, while their transmitted genes mutate and recombine. How, then, are we to proceed with a scientific investigation of evolutionary genetics when we cannot describe that which interests us the most? Population geneticists have achieved remarkable success by choosing to ignore the complexities of real populations and focusing on the evolution of one or a few loci at a time in a population that is assumed to mate at random or, if subdivided, to have a simple migration pattern. The success of this approach, which is seen in both theoretical and experimental investigations, has been impressive, as I hope the reader will agree by the end of this book. The approach is not without its detractors. Years ago, Ernst Mayr mocked this approach as "bean bag genetics." In so doing, he echoed a view held by many of the pioneers of our field that natural selection acts on highly interactive coadapted genomes whose evolution cannot be understood by considering the evolution of a few loci in isolation from all others. Although genomes are certainly coadapted, there is precious little evidence that there are strong interactions between most polymorphic alleles in natural populations. The modern view, spurred on by the rush of DNA sequence data, is that we can profitably study loci in isolation.

This chapter begins with a description of the genetic structure of the alcohol dehydrogenase locus, *ADH*, in *Drosophila*. *ADH* is but one locus in one species. Yet, its genetic structure is typical in most regards. Other loci in *Drosophila* and in other species may differ quantitatively, but not in their gross features.

1.1 DNA variation in Drosophila

Although population genetics is concerned mainly with genetic variation within species, until recently only genetic variation with major morphological manifestations, such as visible, lethal, or chromosomal mutations, could be analyzed genetically. The bulk of genetically based variation was refractory to the most sensitive of experimental protocols. Variation was known to exist because of the uniformly high heritabilities of quantitative traits; there was simply no way to dissect it.

Today, all this has changed. With readily available polymerase chain reaction (PCR) kits, the appropriate primers, and a sequencing machine, even the uninitiated can soon obtain DNA sequences from several alleles in their favorite species. In fact, sequencing is so easy that data are accumulating more rapidly than they can be interpreted.

The 1983 paper "Nucleotide polymorphism at the alcohol dehydrogenase locus of *Drosophila melanogaster*," by Marty Kreitman, was a milestone in evolutionary genetics because it was the first to describe sequence variation in a sample of alleles obtained from nature. At the time, it represented a prodigious amount of work. Today, a mere 13 years later, an undergraduate could complete the study in a few weeks. The alcohol dehydrogenase locus in *D. melanogaster* has the typical exon-intron structure of eukaryotic genes. Only the 768 bases of the coding sequence are given in Figure 1.1, along with its translation.

Kreitman sequenced 11 alleles from Florida (Fl), Washington (Wa), Africa (Af), Japan (Ja), and France (Fr). When the sequences were compared base by base, it turned out that they were not all the same. In fact, no two alleles had exactly the same DNA sequence, although within just the coding sequences, as illustrated in Figure 1.1, some alleles did have the same sequence.

Within the coding region of the 11 *ADH* alleles, 14 sites have two alternative nucleotides. These are listed in Table 1.1 and their positions are illustrated in Figure 1.1. A site with different nucleotides in independently sampled alleles is called a segregating site; less often, it is called a polymorphic site. About 1.8 of every 100 sites are segregating in the *ADH* sample, a figure that is typical for *D. melanogaster* loci. The variation at 13 of the 14 segregating sites is silent, so called because the alternative codons code for the same amino acid. The variation at the 578th nucleotide position results in a change of the amino acid at position 192 in the protein, where either a lysine (AAG) or a threonine (ACG) is found. A nucleotide polymorphism that causes an amino acid polymorphism

```
1                                                 g
atg.tcg.ttt.act.ttg.acc.aac.aag.aac.gtg.att.ttc.gtt.gcc.ggt.ctg.gga.ggc.att.ggt
Met.Ser.Phe.Thr.Leu.Thr.Asn.Lys.Asn.Val.Ile.Phe.Val.Ala.Gly.Leu.Gly.Gly.Ile.Gly
61
ctg.gac.acc.agc.aag.gag.ctg.ctc.aag.cgc.gat.ctg.aag.aac.ctg.gtg.atc.ctc.gac.cgc
Leu.Asp.Thr.Ser.Lys.Glu.Leu.Leu.Lys.Arg.Asp.Leu.Lys.Asn.Leu.Val.Ile.Leu.Asp.Arg
121
att.gag.aac.ccg.gct.gcc.att.gcc.gag.ctg.aag.gca.atc.aat.cca.aag.gtg.acc.gtc.acc
Ile.Glu.Asn.Pro.Ala.Ala.Ile.Ala.Glu.Leu.Lys.Ala.Ile.Asn.Pro.Lys.Val.Thr.Val.Thr
181                                                         t
ttc.tac.ccc.tat.gat.gtg.acc.gtg.ccc.att.gcc.gag.acc.acc.aag.ctg.ctg.aag.acc.atc
Phe.Tyr.Pro.Tyr.Asp.Val.Thr.Val.Pro.Ile.Ala.Glu.Thr.Thr.Lys.Leu.Leu.Lys.Thr.Ile
241
ttc.gcc.cag.ctg.aag.acc.gtc.gat.gtc.ctg.atc.aac.gga.gct.ggt.atc.ctg.gac.gat.cac
Phe.Ala.Gln.Leu.Lys.Thr.Val.Asp.Val.Leu.Ile.Asn.Gly.Ala.Gly.Ile.Leu.Asp.Asp.His
301
cag.atc.gag.cgc.acc.att.gcc.gtc.aac.tac.act.ggc.ctg.gtc.aac.acc.acg.acg.gcc.att
Gln.Ile.Glu.Arg.Thr.Ile.Ala.Val.Asn.Tyr.Thr.Gly.Leu.Val.Asn.Thr.Thr.Thr.Ala.Ile
361                               t       a
ctg.gac.ttc.tgg.gac.aag.cgc.aag.ggc.ggt.ccc.ggt.ggt.atc.atc.tgc.aac.att.gga.tcc
Leu.Asp.Phe.Trp.Asp.Lys.Arg.Lys.Gly.Gly.Pro.Gly.Gly.Ile.Ile.Cys.Asn.Ile.Gly.Ser
421                       a
gtc.act.gga.ttc.aat.gcc.atc.tac.cag.gtg.ccc.gtc.tac.tcc.ggc.acc.aag.gcc.gcc.gtg
Val.Thr.Gly.Phe.Asn.Ala.Ile.Tyr.Gln.Val.Pro.Val.Tyr.Ser.Gly.Thr.Lys.Ala.Ala.Val
481                                       a       c               g           t
gtc.aac.ttc.acc.agc.tcc.ctg.gcg.aaa.ctg.gcc.ccc.att.acc.ggc.gtg.acc.gct.tac.acc
Val.Asn.Phe.Thr.Ser.Ser.Leu.Ala.Lys.Leu.Ala.Pro.Ile.Thr.Gly.Val.Thr.Ala.Tyr.Thr
541                                              c
gtg.aac.ccc.ggc.atc.acc.cgc.acc.acc.ctg.gtg.cac.aag.ttc.aac.tcc.tgg.ttg.gat.gtt
Val.Asn.Pro.Gly.Ile.Thr.Arg.Thr.Thr.Leu.Val.His.Lys.Phe.Asn.Ser.Trp.Leu.Asp.Val
601   t           c                                       c
gag.ccc.cag.gtt.gct.gag.aag.ctc.ctg.gct.cat.ccc.acc.cag.cca.tcg.ttg.gcc.tgc.gcc
Glu.Pro.Gln.Val.Ala.Glu.Lys.Leu.Leu.Ala.His.Pro.Thr.Gln.Pro.Ser.Leu.Ala.Cys.Ala
661                           a
gag.aac.ttc.gtc.aag.gct.atc.gag.ctg.aac.cag.aac.gga.gcc.atc.tgg.aaa.ctg.gac.ctg
Glu.Asn.Phe.Val.Lys.Ala.Ile.Glu.Leu.Asn.Gln.Asn.Gly.Ala.Ile.Trp.Lys.Leu.Asp.Leu
721
ggc.acc.ctg.gag.gcc.atc.cag.tgg.acc.aag.cac.tgg.gac.tcc.ggc.atc.
Gly.Thr.Leu.Glu.Ala.Ile.Gln.Trp.Thr.Lys.His.Trp.Asp.Ser.Gly.Ile.
```

Figure 1.1: The DNA sequence for the coding region of the reference allele from the alcohol dehydrogenase locus of *Drosophila melanogaster*. The translation, given below the DNA sequence, uses the three-letter codes for amino acids. The letters over certain bases indicate the variants for those nucleotides found in a sample from nature. The variant at position 578 changes the amino acid of its codon from lysine to threonine.

Allele	39	226	387	393	441	513	519	531	540	578	606	615	645	684
Reference	T	C	C	C	C	C	T	C	C	A	C	T	A	G
Wa-S	.	T	T	.	A	A	C	.	.	.	.	.	.	.
Fl-1S	.	T	T	.	A	A	C	.	.	.	.	.	.	.
Af-S	.	.	.	.	.	.	.	.	.	.	.	.	.	A
Fr-S	.	.	.	.	.	.	.	.	.	.	.	.	.	A
Fl-2S	G	.	.	.	.	.	.	.	.	.	.	.	.	.
Ja-S	G	.	.	.	.	.	.	.	T	.	T	.	C	A
Fl-F	G	.	.	.	.	.	.	G	T	C	T	C	C	.
Fr-F	G	.	.	.	.	.	.	G	T	C	T	C	C	.
Wa-F	G	.	.	.	.	.	.	G	T	C	T	C	C	.
Af-F	G	.	.	.	.	.	.	G	T	C	T	C	C	.
Ja-F	G	.	.	A	.	.	.	G	T	C	T	C	C	.

Table 1.1: The 11 *ADH* alleles. A dot is placed when a nucleotide is the same as the nucleotide in the reference sequence. The numbers refer to the position in the coding sequence where the 14 variant nucleotides are found (see Figure 1.1). The first two letters of the allele name identify the place of origin. The S alleles have a lysine at position 192 of the protein; the F alleles have a threonine.

is called a replacement polymorphism.*

Kreitman's data pose a question which is the Great Obsession of population geneticists: What evolutionary forces could have led to such divergence between individuals within the same species? A related question that sheds light on the Great Obsession is: Why the preponderance of silent over replacement polymorphisms? The latter question is more compelling when you consider that about three-quarters of random changes in a typical DNA sequence will cause an amino acid change. Rather than 75 percent of the segregating sites being replacement, only 7 percent are replacement. Perhaps silent variation is more common because it has a very small effect on the phenotype. By contrast, a change in a protein could radically alter its function. Alcohol dehydrogenase is an important enzyme because flies and their larvae are often found in fermenting fruits with a high alcohol concentration. Inasmuch as alcohol dehydrogenase plays a role in the detoxification of ingested alcohol, a small change in the protein could have substantial physiologic consequences. Thus, it is reasonable to suggest that selection on amino acid variation in proteins will be stronger than on silent variation and that the stronger selection might reduce the level of polymorphism. This is a good suggestion, but it is only a suggestion. Population geneticists take such suggestions and turn them into testable scientific hypotheses, as will be seen as this book unfolds.

Just as there is *ADH* variation within species, so too is there variation between species, as illustrated in Figure 1.2. In this figure, the coding region of the *ADH* locus in *D. melanogaster* is compared to that of the closely related

*Some people use *synonymous* and *nonsynonymous* as synonyms for *silent* and *replacement*, respectively.

species, *D. erecta*. Thirty-six of 768 nucleotides differ between the two species. The probability that a randomly chosen site is different is 36/768 = 0.0468; note that this is also the average number of nucleotide differences per site. Of the 36 differences, only 10 (26%) result in amino acid differences between the two species. Kreitman's polymorphism data also exhibited less replacement than silent variation, but the disparity was somewhat greater: one replacement difference out of 14 (7%) segregating sites.

The comparison of variation within and between species shows no striking lack of congruence. In both cases, all of the differences involve only isolated nucleotides and, in both cases, there are more silent than replacement changes. Things could have been otherwise. For example, the variation within species could have involved isolated nucleotide changes while the differences between species could have been due to insertions and deletions. Were this observed, then the variation within species would have little to contribute to our understanding of evolution in the broader sense. As it is, population geneticists feel confident that their studies of variation within populations play a key role in the wider discipline of evolutionary biology.

Molecular variation may seem far removed from what interests most evolutionists. For many, the allure of evolution is the understanding of the processes leading to the strange creatures of the past or the sublime adaptations of modern species. The raw material of this evolution, however, is just the sort of molecular variation described above. Later in the book, we will be examining genetic variation in fitness traits, as illustrated in Figure 3.6, and in quantitative traits, as illustrated in Figure 5.1. This genetically determined variation must ultimately be due to the kind of molecular variation observed at the *ADH* locus. As of this writing, the connections between molecular variation and phenotypic variation have not been made. The discovery of these connections remains one of the great frontiers of population genetics. Of particular interest in this endeavor will be the relative roles played by variation in coding regions, as seen in the *ADH* example, and variation in the control regions just upstream from coding regions.

1.2 Loci and alleles

We must now make a short digression into vocabulary because two words, *locus* and *allele*, must be made more precise than is usual in genetics textbooks. Although the terms were used without ambiguity for many years, the increase in our understanding of molecular genetics has clouded their original meanings considerably. Here we will use *locus* to refer to the place on a chromosome where an allele resides. An *allele* is just the bit of DNA at that place. A locus is a template for an allele. An allele is an instantiation of a locus. A locus is not a tangible thing; rather, it is a map describing where to find a tangible thing, an allele, on a chromosome. (Some books use *gene* as a synonym for our *allele*. However, *gene* has been used in so many different contexts that it is not very useful for our purposes.) With this convention, a diploid individual may be said

```
atg.tcg.ttt.act.ttg.acc.aac.aag.aac.gtg.att.ttc.gtt.gcc.ggt.ctg.gga.ggc.att.ggt
   .g a.  c.  c.   .   .   .   .   .  c.   .   .  g.   .   .   .   .   .   .  c
   .Ala.   .   .   .   .   .   .   .   .   .   .   .   .   .   .   .   .   .

ctg.gac.acc.agc.aag.gag.ctg.ctc.aag.cgc.gat.ctg.aag.aac.ctg.gtg.atc.ctc.gac.cgc
   .   .   .   .   .   .   .g  .  t.   .   .   .   .   .   .   .   .   .   .
   .   .   .   .   .   .   .Val.   .   .   .   .   .   .   .   .   .   .   .

att.gag.aac.ccg.gct.gcc.att.gcc.gag.ctg.aag.gca.atc.aat.cca.aag.gtg.acc.gtc.acc
   .   .   .   .  c.   .   .   .   .   .   .   .   .   .   .   .   .   .   .
   .   .   .   .   .   .   .   .   .   .   .   .   .   .   .   .   .   .   .

ttc.tac.ccc.tat.gat.gtg.acc.gtg.ccc.att.gcc.gag.acc.acc.aag.ctg.ctg.aag.acc.atc
   .  t.   .   .   .   .   .   .   .   .   .   .   . g .   .   .  c.   .   .
   .   .   .   .   .   .   .   .   .   .   .   .   .Ser.   .   .   .   .   .

ttc.gcc.cag.ctg.aag.acc.gtc.gat.gtc.ctg.atc.aac.gga.gct.ggt.atc.ctg.gac.gat.cac
   .   .a  .   . c .   .   .   .   .   .   .   .   .   .   .   .   .   .   .t
   .   .Lys.   .Thr.   .   .   .   .   .   .   .   .   .   .   .   .   .   .Tyr

cag.atc.gag.cgc.acc.att.gcc.gtc.aac.tac.act.ggc.ctg.gtc.aac.acc.acg.acg.gcc.att
   .   .   .   .   .   .   .   .   .   .   .   .   .   .   .   .   .   .   .
   .   .   .   .   .   .   .   .   .   .   .   .   .   .   .   .   .   .   .

ctg.gac.ttc.tgg.gac.aag.cgc.aag.ggc.ggt.ccc.ggt.ggt.atc.atc.tgc.aac.att.gga.tcc
   .   .   .   .   .   .   .   .   .   .   .   .  c.  t.   .   .   .   .   .
   .   .   .   .   .   .   .   .   .   .   .   .   .   .   .   .   .   .   .

gtc.act.gga.ttc.aat.gcc.atc.tac.cag.gtg.ccc.gtc.tac.tcc.ggc.acc.aag.gcc.gcc.gtg
  g.   .   .   .   .   .   .   .   .   .   .   .   .  t.   .   .   .  t.   .
   .   .   .   .   .   .   .   .   .   .   .   .   .   .   .   .   .   .   .

gtc.aac.ttc.acc.agc.tcc.ctg.gcg.aaa.ctg.gcc.ccc.att.acc.ggc.gtg.acc.gct.tac.acc
   .   .   .   .   .   .   .   .   .   .   .   .  c.   .   .   .   .   .  t.
   .   .   .   .   .   .   .   .   .   .   .   .   .   .   .   .   .   .   .

gtg.aac.ccc.ggc.atc.acc.cgc.acc.acc.ctg.gtg.cac.aag.ttc.aac.tcc.tgg.ttg.gat.gtt
   .   .   .   .   .   .   .   .   .   .   .   .   .   .   .   .   .c  .   .
   .   .   .   .   .   .   .   .   .   .   .   .   .   .   .   .   .   .   .

gag.ccc.cag.gtt.gct.gag.aag.ctc.ctg.gct.cat.ccc.acc.cag.cca.tcg.ttg.gcc.tgc.gcc
   .   .   .  g.  c.   .   .   .   .   .   .   .   .   .a c.   .   .t  .   .
   .   .   .   .   .   .   .   .   .   .   .   .   .   .Thr.   .   .Ser.   .

gag.aac.ttc.gtc.aag.gct.atc.gaa.ctg.aac.cag.aac.gga.gcc.atc.tgg.aaa.ctg.gac.ctg
   .   .  t.   .   .  c.   .  g.   .   .g      .  t.   .   .   .   .   .   .
   .   .   .   .   .   .   .   .   .   .Glu.   .   .   .   .   .   .   .   .

ggc.acc.ctg.gag.gcc.atc.cag.tgg.acc.aag.cac.tgg.gac.tcc.ggc.atc.
   .   .   .   .   .   .  a.   . g .   .   .   .   .   .   .   .
   .   .   .   .   .   .   .   .Ser.   .   .   .   .   .   .   .
```

Figure 1.2: The DNA sequence for *D. melanogaster* ADH with those bases and amino acids that differ in *D. erecta* shown below. The *erecta* sequence is from Jeffs et al. (1994).

to have two alleles at a particular autosomal locus, one from its mother and the other from its father.

Population genetics, like other areas of genetics, is concerned with alleles that differ one from another. However, in population genetics there are subtleties in what is meant by "different alleles." There are three fundamental ways in which alleles at the same locus may differ:

By origin. Alleles differ by origin if they come from the same locus on different chromosomes. One often refers to a sample of n (different) alleles from a population. What is meant by "different" in this context is "different by origin." For example, the two alleles at a specified locus in a diploid individual are always different by origin. The 11 alleles in Kreitman's sample also differ by origin.

By state. Whether or not two alleles are said to differ by state depends on the context. If the context is the DNA sequence of the alleles, then they are different by state if they have different DNA sequences. The difference may as small as one nucleotide out of thousands. However, in evolutionary studies we frequently focus on particular aspects of alleles and may choose to put them in different states depending on the nature of the difference. For example, if our interest is in protein evolution, we may choose to say that two alleles are different by state if and only if they differ in their amino acid sequences. (We do this in full recognition that some alleles with the same amino acid sequence may have different DNA sequences as a consequence of the redundancy of the genetic code.) Similarly, we may choose to call two alleles different by state if and only if they have different amino acids at a particular site, perhaps at the fourth position in the protein. States may also be thought of as phenotypes, which could include the DNA sequence, the protein sequence, the color of the pea, or other genetically determined phenotypes of interest.

By descent. Alleles differ by descent when they do not share a common ancestor allele. Strictly speaking, two alleles from the same locus can never be different by descent as all contemporary alleles share a remote common ancestor. In practice, we are often concerned with a relatively short time in the past and are content to say that two alleles are different by descent if they do not share a common ancestor allele in, say, the past 10 generations. Two alleles that are different by descent may or may not be different by state because of mutation. Difference by descent will not be used until Section 4.2.

The converse of the above involves identity by origin, state, or descent.Alleles that are identical by origin are necessarily identical by state and descent. Two alleles that are identical by descent may not be identical by state because of mutation. Figure 1.3 gives a simple example of three nucleotides in alleles obtained from two individuals in generation n and traced back to their ancestor allele in generation $n-2$. The two alleles are identical by descent because they

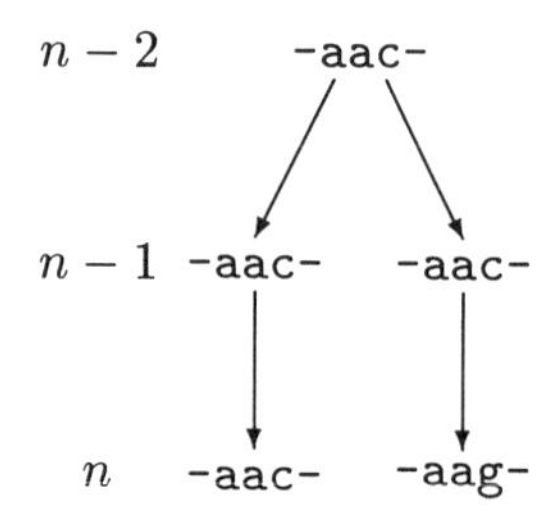

Figure 1.3: Two alleles in generation n that are identical by descent but differ in state.

are both copies of the same ancestor allele in the recent past. However, they are different by state because a mutation from c to g appeared in the right-hand allele.

Diploid individuals are said to be heterozygous at a locus if the two alleles at that locus are different by state. They are homozygous if their two alleles are identical by state. The use of homozygous or heterozygous is always in the context of the states under study. If we are studying proteins, we may call an individual homozygous at a locus when the protein sequences of the two alleles are identical, even if their DNA sequences are different.

Originally, *alleles* referred to different states of a gene. Our definition differs from this traditional usage in that alleles exist even if there is no genetic variation at a locus. Difference by origin has not been used before. It is introduced here to be able to use phrases like "a sample of n different alleles" without implying that the alleles are different by state.

Kreitman's sample contains 11 alleles that differ by origin. How many alleles differ by state? If we were interested in the full DNA sequence, then the sample contains six alleles that are different by state. If we were interested in proteins, then the sample contains only two alleles that differ by state. Of the two protein alleles, the one with a lysine at position 192 makes up $6/11 = 0.55$ of the alleles. The usual way to say this is that the allele frequency of the lysine-containing allele in the sample is 0.55. The sample allele frequency is an estimate of the population allele frequency. It's not a particularly precise estimate because of the small sample size. A rough approximation to the 95 percent confidence interval for a proportion is

$$\hat{p} \pm 1.96\sqrt{\hat{p}(1-\hat{p})/n},$$

where $\hat{p}$ is the estimate of the proportion, 0.55 in our case, and n is the sample size. Thus, the probability that the population allele frequency falls within the interval $(0.25, 0.78)$ is 0.95. If a more precise estimate is needed, the sample size would have to be increased.

1.3 Genotype and allele frequencies

Population genetics is very quantitative. A description of the genetic structure of a population is seldom simply a list of genotypes, but rather uses relative frequencies of alleles and genotypes. With quantification comes a certain degree of abstraction. For example, to introduce the notion of genotype and allele frequencies we will not refer to a particular sample, like Kreitman's *ADH* sample, but rather to a locus that we will simply call the *A* locus. (No harm will come in imagining the *A* locus to be the *ADH* locus.) Initially, we will assume that the locus has two alleles, called A_1 and A_2, segregating in the population. (These could be the two protein alleles at the *ADH* locus.) By implication, these two alleles are different by state. There will be three genotypes in the population: two homozygous genotypes, A_1A_1 and A_2A_2, and one heterozygous genotype, A_1A_2. The relative frequency of a genotype will be written x_{ij}, as illustrated in the following table.

Genotype:	A_1A_1	A_1A_2	A_2A_2
Relative frequency:	x_{11}	x_{12}	x_{22}

As the relative frequencies must add to one, we have

$$x_{11} + x_{12} + x_{22} = 1.$$

The ordering of the subscripts for heterozygotes is arbitrary. We could have used x_{21} instead of x_{12}. However, it is not permissible to use both. In this book, we will always use the convention of making the left index the numerically smaller one.

Allele frequencies play as important a role in population genetics as do genotype frequencies. The frequency of the A_1 allele in the population is

$$p = x_{11} + \frac{1}{2}x_{12}, \tag{1.1}$$

and the frequency of the A_2 allele is

$$q = 1 - p = x_{22} + \frac{1}{2}x_{12}.$$

We can think of the allele frequency, p, in two different ways. One is simply as the relative frequency of A_1 alleles among all of the A alleles in the population. The other is as the probability that an allele picked at random from the population is an A_1 allele. The act of picking an allele at random may be broken down into a sequence of two actions: picking a genotype at random from the population and then picking an allele at random from the chosen genotype. Because there are three genotypes, we could write p as

$$p = (x_{11} \times 1) + (x_{12} \times \frac{1}{2}) + (x_{22} \times 0).$$

This representation shows that there are three mutually exclusive ways in which we might obtain an A_1 allele and gives the probability of each. For example, the first term in the sum is the joint event that an A_1A_1 is chosen (this occurs with probability x_{11}) and that an A_1 allele is subsequently chosen from the A_1A_1 individual (this occurs with probability one). It is difficult to underestimate the importance of probabilistic reasoning when doing population genetics. I urge the reader to think carefully about the probabilistic definition of p until it becomes second nature.

Most loci have more than two alleles. In such cases, the frequency of the ith allele will be called p_i. As before, the frequency of the A_iA_j genotype will be called x_{ij}. For heterozygotes, $i \neq j$ and, by convention, $i < j$. As with the two-allele case, the sum of all of the genotype frequencies must add to one. For example, if there are n alleles, then

$$\begin{aligned} 1 &= x_{11} + x_{22} + \cdots + x_{nn} + x_{12} + x_{13} + \cdots + x_{(n-1)n} \\ &= \sum_{i=1}^{n} \sum_{j \geq i}^{n} x_{ij}. \end{aligned}$$

The frequency of the ith allele is

$$p_i = x_{ii} + \frac{1}{2} \sum_{j=1}^{i-1} x_{ji} + \frac{1}{2} \sum_{j=i+1}^{n} x_{ij}$$

Again, this allele frequency has both a relative frequency and a probabilistic interpretation.

Problem 1.1 *How many different genotypes are there at a locus with n alleles that differ by state? You already know that there is one genotype at a locus with one allele and three genotypes at a locus with two alleles. Continue this with three, four, and more alleles until you divine the general case. (The answers to select problems, including this one, are found at the end of each chapter.)*

In the mid-1960s, population geneticists began to use electrophoresis to describe genetic variation in proteins. For the first time, the genetic variation at a "typical" locus could be ascertained. Harry Harris's 1966 paper, "Enzyme polymorphism in man," was among the first of many electrophoretic survey papers. In it, he summarized the electrophoretic variation at 10 loci sampled from the English population. The protein produced by one of these loci is placental alkaline phosphatase. Harris found three phosphatase alleles that differed by state (migration speed) and called them S (slow), I (intermediate), and F (fast) for their rate of movement in the electrophoresis apparatus. The genotype frequencies are given in Table 1.2.

The frequency of heterozygotes at the placental alkaline phosphatase locus is $158/332 = 0.48$, which is unusually high for human protein loci. The average probability that an individual is heterozygote at a locus examined in this paper is approximately 0.05. If this could be extrapolated to the entire genome, then a typical individual would be heterozygous at 1 (at least) of every 20 loci. However, there is evidence that the enzymes used in Harris's study are not "typical"

Genotype	Number	Frequency	Expected
SS	141	0.4247	0.4096
SF	111	0.3343	0.3507
FF	28	0.0843	0.0751
SI	32	0.0964	0.1101
FI	15	0.0452	0.0471
II	5	0.0151	0.0074
Total	332	1.0000	1.0000

Table 1.2: The frequencies of alkaline phosphatase genotypes in a sample from the English people. The expected Hardy-Weinberg frequencies are given in the fourth column. The data are from Harris (1966).

loci. They appear to be more variable than other protein loci. At present, we do not have a reliable estimate of the distribution of protein heterozygosities across loci for any species.

Problem 1.2 *Calculate the frequency of the three alkaline phosphatase alleles in the English population.*

1.4 Randomly mating populations

The first milestone in theoretical population genetics, the celebrated Hardy-Weinberg law, was the discovery of a simple relationship between allele frequencies and genotype frequencies at an autosomal locus in an equilibrium randomly mating population. That such a relationship might exist is suggested by the pattern of genotype frequencies in Table 1.2. For example, the S allele is more frequent than the F allele and the SS homozygote is more frequent than the FF homozygote, suggesting that homozygotes of more frequent alleles will be more common than homozygotes of less frequent alleles. Such qualitative observations yield quite naturally to the desire for quantitative relationships between allele and genotype frequencies, as provided by the insights of George Hardy and Wilhelm Weinberg.

The Hardy-Weinberg law describes the equilibrium state of a single locus in a randomly mating diploid population that is free of other evolutionary forces, such as mutation, migration, and genetic drift. By random mating, we mean that mates are chosen with complete ignorance of their genotype (at the locus under consideration), degree of relationship, or geographic locality. For example, a population in which individuals prefer to mate with cousins is not a randomly mating population. Rather, it is an inbreeding population. A population in which A_1A_1 individuals prefer to mate with other A_1A_1 individuals is not a randomly mating population either. Rather, this population is experiencing assortative mating. Geography can also prevent random mating if individuals are more likely to mate with neighbors than with mates chosen at random from the entire species. Inbreeding and population subdivision will be examined in

Chapter 4. Assortative mating will not be discussed further because it is a specialized topic, although one that can play an important role in the evolution of some species.

The Hardy-Weinberg law is particularly easy to understand in hermaphroditic species (species in which each individual is both male and female). The autosomal loci of hermaphrodites reach their Hardy-Weinberg equilibrium in a single generation of random mating, no matter how far the initial genotype frequencies are from their equilibrium values. Our task, then, is to study the change in genotype frequencies in hermaphrodites brought about by random mating at an autosomal locus with two alleles, A_1 and A_2, and genotype frequencies x_{11}, x_{12}, and x_{22}.

To form a zygote in the offspring generation, the assumption of random mating requires that we choose two gametes at random from the parent generation. The probability that the zygote is an A_1A_1 homozygote is the product of the probability that the egg is A_1, p, times the probability that the sperm is A_1, also p. (The fact that these two probabilities are the same is the consequence of assuming that the species is hermaphroditic.) Thus, the probability that a randomly formed zygote is A_1A_1 is just p^2 by the product rule of probabilities for independent events. Similarly, the probability that a randomly formed zygote is A_2A_2 is q^2. An A_1A_2 heterozygote may be formed in two different ways. One way is with an A_1 egg and an A_2 sperm. The probability of this combination is pq. The other way is with an A_2 egg and an A_1 sperm. The probability of this combination is also pq. Thus, the total probability of forming a heterozygote is $2pq$ by the addition rule of probabilities for mutually exclusive events.

After one round of random mating, the frequencies of the three genotypes are

Genotype:	A_1A_1	A_1A_2	A_2A_2
Frequency (H-W):	p^2	$2pq$	q^2

These are the Hardy-Weinberg genotype frequencies. As advertised, they depend only on the allele frequencies: If you know p, then you know the frequencies of all three genotypes.

The important things to note about the evolutionary change brought about by random mating in diploid hermaphroditic populations are:

- The frequencies of the alleles do not change as a result of random mating, as may be seen by using Equation 1.1 with the Hardy-Weinberg frequencies. Random mating can change genotype frequencies, not allele frequencies. Consequently, the Hardy-Weinberg genotype frequencies will remain unchanged in all generations after the first.

- The Hardy-Weinberg equilibrium is attained in only one round of random mating. This is traceable to our assumption that the species is hermaphroditic (and that we are studying an autosomal locus). In a species with separate sexes, it takes two generations to achieve Hardy-Weinberg equilibrium, as we will soon discover.

- To calculate the genotype frequencies after a round of random mating, we need only the allele frequencies before random mating, not the genotype frequencies.

Of course, many species are not hermaphrodites but are dioecious; each individual is either male or female. To further complicate matters, the genotype frequencies could be different in the two sexes. As an extreme example, suppose that all of the females are A_1A_1 and all of the males are A_2A_2. If the sexes are equally frequent, the frequency of the A_1 allele in the population is $p = 1/2$. After one round of random mating, the frequencies of the A_1A_1 and A_2A_2 homozygotes are zero, and the frequency of the A_1A_2 heterozygote is one. These frequencies are far from the Hardy-Weinberg frequencies. However, the third generation, produced by random mating of heterozygotes, has genotypes A_1A_1, A_1A_2, and A_2A_2 in the Hardy-Weinberg frequencies 1/4, 1/2, and 1/4, respectively. Thus, for dioecious species with unequal genotype frequencies in the two sexes, it can take two generations to reach equilibrium.

Can it take more or fewer? The answer depends on whether the locus is on an autosome or a sex chromosome. For now, consider only the case of an autosomal locus, for which one round of random mating makes the allele frequencies the same in both sexes and equal to the average of the frequencies in the males and females of the parent or first generation. Call the frequency of the A_1 allele in the first and second generations p.

In the next generation (the third), the probability that a zygote is A_1A_1 is the product of the probabilities that the sperm is A_1, p, and that the egg is A_1, also p. These two probabilities became equal in the second generation. From here on, the argument parallels that used for hermaphrodites with the same ultimate genotype frequencies. Thus, if the allele frequencies are different in the two sexes, it takes two generations to reach Hardy-Weinberg frequencies. Otherwise, it takes only one generation.

Problem 1.3 *Hardy-Weinberg frequencies in dioecious species may be investigated in an entirely different way. Let the genotype frequencies in females be x_{11}, x_{12}, and x_{22}, and in males, y_{11}, y_{12}, and y_{22}. Enumerate all nine possible matings (A_1A_1 female by A_1A_1 male, A_1A_1 female by A_1A_2 male, etc.) and calculate the frequencies of genotypes produced by each one as a function of the x's and y's. Sum these frequencies, weighted by the frequencies of the matings, to obtain the genotype frequencies in the second generation. Now let these genotypes mate at random to produce the third generation. If all goes well, a morass of symbols will collapse into the satisfying simplicity of the Hardy-Weinberg law. Be forewarned, you will need several sheets of paper.*

Problem 1.4 *Graph the frequencies of homozygotes and heterozygotes as a function of the allele frequency, p. At what allele frequency is the frequency of heterozygotes maximized?*

One of the most important consequences of the Hardy-Weinberg law concerns the genotypes occupied by rare alleles. Suppose the A_2 allele is rare; that is,

$q = 1 - p$ is small. Are A_2 alleles more likely to be in A_2A_2 homozygotes or A_1A_2 heterozygotes? The ratio of the latter to the former is

$$\frac{2pq}{q^2} = \frac{2p}{q} \approx \frac{2}{q}.$$

The approximation used in the last step makes use of the assumption that q is small. As $p = 1 - q$, p may be approximated by one because q is small relative to one. For example, if q is about 0.01, the error in this approximation is about 1 percent, which is perfectly acceptable for population genetics. If $q = 0.01$, an A_2 allele is about 200 times more likely to be in a heterozygote than in a homozygote. If $q = 0.001$, it is about 2000 times more likely to be in a heterozygote. Thus, rare alleles mostly find themselves in heterozygotes and, as a consequence, their fate is tied to their dominance relationship with the A_1 allele. This is our first clue that dominance is an important factor in evolution.

Problem 1.5 *Graph the ratio of the frequencies of A_1A_2 heterozygotes to A_2A_2 homozygotes as a function of q using both the exact and the approximate formulae.*

The generalization of the Hardy-Weinberg law to multiple alleles requires no new ideas. Let the frequency of the k alleles, $A_i, i = 1 \ldots k$, be $p_i, i = 1 \ldots k$. Using the same argument as before, it should be clear that the frequency of the A_iA_i homozygote after random mating will be p_i^2 and the frequency of the A_iA_j heterozygote will be $2p_ip_j$. The total frequency of homozygotes is given by

$$G = \sum_{i=1}^{k} p_i^2. \tag{1.2}$$

G is called the homozygosity of the locus. The heterozygosity of the locus is given by

$$H = 1 - G = 1 - \sum_{i=1}^{k} p_i^2.$$

For randomly mating diploid populations, the heterozygosity equals the frequency of heterozygotes. Note, however, that the definition of heterozygosity uses only allele frequencies, not genotype frequencies. Because of this, heterozygosity is often used to describe levels of variation in populations that do not conform to the Hardy-Weinberg assumption of random mating. It is even used to describe variation in bacterial populations, which are haploids.

The frequencies of the S, F, and I alleles of placental alkaline phosphatase as obtained from Table 1.2 are 0.640, 0.274, and 0.086, respectively. From these we can calculate the expected frequencies of each genotype in the population assuming that it is in Hardy-Weinberg equilibrium. The expected frequency of SS homozygotes, for example, is $0.64^2 = 0.4095$. The fourth column in Table 1.2 gives the expected frequencies for the remainder of the genotypes. The agreement between the observed and expected numbers is quite good. A

Place	S	F	I	Others	n
England	.637	.270	.085	.008	597
Italy	.661	.256	.075	.007	273
West India	.701	.217	.066	.016	208
Thailand	.746	.081	.165	.008	188
Japan	.724	.038	.236	.003	294
Nigeria	.942	.019	.039		130
Canadian Inuits	.556	.142	.296	.006	81
Papua New Guinea	.880	.050	.068	.002	338

Table 1.3: Geographic variation in the three alleles of placental alkaline phosphatase in humans. The "Others" category is the total frequency of alleles other than the three common alleles. The final column is the sample size. The data are from Roychoudhury and Nei (1988).

χ^2 test does not allow rejection of the Hardy-Weinberg hypothesis at the 5 percent level. The human placental alkaline phosphatase story is typical of many proteins examined from populations that are thought to mate at random. Very few cases of significant departures from Hardy-Weinberg expectations have been recorded in outbreeding species.

Problem 1.6 *Derive the Hardy-Weinberg law for a sex-linked locus. Let the initial frequency of A_1 in females be p_f and in males, p_m. Follow the two allele frequencies in successive generations until you understand the allele-frequency dynamics. Then, jump ahead and find the equilibrium genotype frequencies in females and males. Finally, graph the male and female allele frequencies over several generations for a population that is started with all A_1A_1 females ($p_f = 1$) and A_2 males ($p_m = 0$).*

A description of the genetic structure of a population must include a geographic component if the ultimate goal is to understand the evolutionary forces responsible for genetic variation. A conjecture about the evolutionary history of the alkaline phosphatase alleles, for example, will be of one sort if the allele frequencies are the same in all subpopulations and of quite another sort if the subpopulations vary in their allele frequencies. Some representative frequencies taken from the 1988 compilation of human polymorphism data by Arun Roychoudhury and Masatoshi Nei are given in Table 1.3. As is apparent, there is considerable geographic variation in the frequencies of the three alleles. We must conclude what will be obvious to most: The human population is not one large, randomly mating population. The agreement of the genotype frequencies with Hardy-Weinberg expectations within the English population, however, suggests that local groupings have historically approximated randomly mating populations. Most other species show a similar pattern. Some have less differentiation between geographic areas, others quite a bit more. But most show some differentiation, and this fact should be incorporated into our view of the genetic

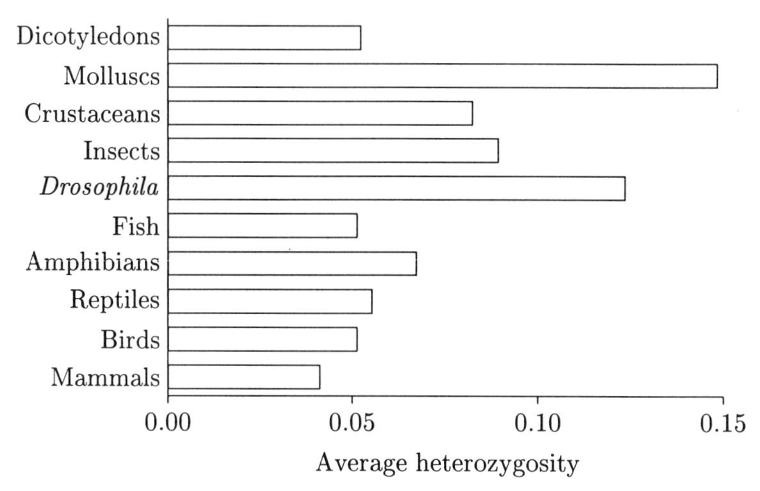

Figure 1.4: Estimates of average protein heterozygosities based on electrophoretic studies plotted from the data in Nevo et al. (1984).

structure of populations. We return to the analysis of data from subdivided populations in Section 4.4.

A great deal of work has been done to estimate the levels of genetic variation in natural populations. For electrophoretically detectable protein variation, average heterozygosities (averaged across loci) vary from zero to about 0.15. The average protein heterozygosity in humans is about 0.05; for *D. melanogaster* it is about 0.12. Figure 1.4 shows that average heterozygosities vary only by about 3.5-fold over a taxonomically diverse group of animals. Average heterozygosities are somewhat misleading because they bury the fact that there is a tremendous amount of variation between loci in levels of polymorphism. For example, soluble enzymes are much more polymorphic than are abundant nonenzymatic proteins. Nonetheless, average heterozygosities do give the correct impression that there is a lot of genetic variation in natural populations.

Why does the Hardy-Weinberg law play such a central role in population genetics? Consider that life first appeared on Earth about four billion years ago. For the next two billion or so years, the earth was populated by haploid prokaryotes, not diploid eukaryotes. During this time, most of the basic elements of living systems appeared: the genetic code, enzymes, biochemical pathways, photosynthesis, bipolar membranes, structural proteins, and on and on. Thus, the most fundamental innovations evolved in populations where the Hardy-Weinberg law is irrelevant! If population genetics were primarily concerned with the genetic basis of evolution, then it is odd that the Hardy-Weinberg law is introduced so early in most texts (including this one). One might expect to find a development targeted at those important first two billion years with a coda to handle the diploid upstarts.

Like so much of science, the development of population genetics is anthrocentric. Our most consuming interest is with ourselves. In fact, the Weinberg of

Hardy-Weinberg was a human geneticist struggling with the study of inheritance in a species in which setting up informative crosses is frowned upon. *Drosophila* ranks a close second to *Homo* in the eyes of geneticists. Both humans and fruit flies exhibit genetic variation in natural populations, and this variation demands an evolutionary investigation. Population genetics and its Great Obsession grew out of this fascination with variation in species we love, not out of a desire to explain the origin of major evolutionary novelties.

1.5 Answers to problems

1.1 When there are n alleles, there are n homozygous genotypes, $A_iA_i, i = 1\ldots n$. If we first view an A_iA_j heterozygote as distinct from an A_jA_i heterozygote, there are $n(n-1)$ such heterozygotes. The actual number of heterozygotes will be one-half this number, or $n(n-1)/2$. Thus, the total number of genotypes is $n + n(n-1)/2 = n(n+1)/2$.

1.3 The matings and the frequency of genotypes from each mating may be summarized in a table with nine rows, the first three of which are given.

Mating female × male	Frequency	Offspring A_1A_1	A_1A_2	A_2A_2
$A_1A_1 \times A_1A_1$	$x_{11}y_{11}$	1		
$A_1A_1 \times A_1A_2$	$x_{11}y_{12}$	1/2	1/2	
$A_1A_1 \times A_2A_2$	$x_{11}y_{22}$			1

Using the complete table, the frequency of A_1A_1 offspring is

$$\begin{aligned} x_{11}y_{11} &+ x_{11}y_{12}/2 + x_{12}y_{11}/2 + x_{12}y_{12}/4 \\ &= (x_{11} + x_{12}/2)(y_{11} + y_{12}/2) \\ &= p_f p_m, \end{aligned}$$

where p_f and p_m are the frequencies of A_1 in females and males in the original population. Similarly, the frequency of A_1A_2 is $p_f q_m + q_f p_m$ and that of A_2A_2 is $q_f q_m$. As the A locus is autosomal and segregates independent of the sex chromosomes, the frequencies of the three genotypes will be the same in both males and females. The frequency of the A_1 allele in the offspring is thus

$$p = p_f p_m + (p_f q_m + q_f p_m)/2 = (p_f + p_m)/2,$$

which is also the same in both males and females. When these offspring mate among themselves to produce the third generation, the above calculations may be used again, but this time with $p_f = p_m = p$. Thus, the frequency of A_1A_1 in the third generation is $p_f p_m = p^2$, which is the Hardy-Weinberg frequency. The other two genotype frequencies are obtained in a similar way.

1.6 As males get their X-chromosomes from their mothers, the frequency of A_1 in males is always equal to the frequency in females in the previous generation. As a female gets one X from her mother and one from her father, the allele frequency in females is always the average of the male and female frequencies in the previous generation. Thus, the allele frequencies over the first three generations are as follows.

Generation	Females	Males	Female − male
1	p_f	p_m	$p_f - p_m$
2	$(p_f + p_m)/2$	p_f	$-(p_f - p_m)/2$
3	$[p_f + (p_f + p_m)/2]/2$	$(p_f + p_m)/2$	$(p_f - p_m)/4$

Two important things emerge from this table. First, the overall allele frequency,

$$p = \frac{2}{3}p_f + \frac{1}{3}p_m,$$

does not change over time. (Convince yourself that this is so by calculating p in generations 2 and 3.) Second, the difference between the allele frequencies in females and males is halved each generation, as recorded in the table. Taken together, these two observations show that eventually the allele frequencies in males and females will converge to p. At that time, the genotype frequencies in females will be Hardy-Weinberg frequencies.

Chapter 2

Genetic Drift

The discussion of random mating and the Hardy-Weinberg law in the previous chapter was premised on the population size being infinite. Sometimes real populations are very large (roughly 10^9 for our own species), in which case the infinite assumption might seem reasonable, at least as a first approximation. However, the population sizes of many species are not very large. Bird watchers will tell you, for example, that there are fewer than 100 Bachman's warblers in the cypress swamps of South Carolina. For these warblers, the infinite population size assumption of the Hardy-Weinberg law may be hard to accept. In finite populations, random changes in allele frequencies result from variation in the number of offspring between individuals and, if the species is diploid and sexual, from Mendel's law of segregation.

Genetic drift, the name given to these random changes, affects evolution in two important ways. One is as a dispersive force that removes genetic variation from populations. The rate of removal is inversely proportional to the population size, so genetic drift is a very weak dispersive force in most natural populations. The other is drift's effect on the probability of survival of new mutations, an effect that is important even in the largest of populations. In fact, we will see that the survival probability of beneficial mutations is (approximately) independent of the population size.

The dispersive aspect of genetic drift is countered by mutation, which puts variation back into populations. We will show how these two forces reach an equilibrium and how they can account for much of the molecular variation described in the previous chapter.

The neutral theory states that much of molecular variation is due to the interaction of drift and mutation. This theory, one of the great accomplishments of population genetics because it is the first fully developed theory to satisfy the Great Obsession, has remained controversial partly because it has been difficult to test and partly because of its seemingly outrageous claim that most of evolution is due to genetic drift rather than natural selection, as Darwin imagined. The theory will be developed in this chapter and will reappear in several later chapters as we master additional topics relevant to the theory.

2.1 A computer simulation

Simple computer simulations, as shown in Figure 2.1, may be used to illustrate the consequences of genetic drift. These particular simulations model a population of $N = 20$ diploid individuals with two segregating alleles, A_1 and A_2. The frequency of A_1 at the start of each simulation is $p = 0.2$, which represents 8 A_1 alleles and 32 A_2 alleles. Each new generation is obtained from the previous generation by repeating the following three steps $2N = 40$ times.

1. Choose an allele at random from among the $2N$ alleles in the parent generation.

2. Make an exact copy of the allele.

3. Place the copy in the new generation.

After 40 cycles through the algorithm, a new population is created with an allele frequency that will, in general, be different from that of the original population. The reason for the difference is the randomness introduced in step 1.

As written, these steps may be simulated on a computer or with a bag of marbles of two colors, initially 8 of one color and 32 of another (providing that you have all of your marbles). The results of five independent simulations are illustrated in Figure 2.1. Obviously, allele frequencies do change at random. Nothing could be farther from the constancy promised by Hardy-Weinberg.

In natural populations, there are two main sources of randomness. One is Mendel's law of segregation. When a parent produces a gamete, each of its two homologous alleles is equally likely to appear in the gamete. The second is demographic stochasticity.* Different individuals have different numbers of offspring for complex reasons that collectively appear to be random. Neither of these sources gives any preference to particular alleles. Each of the $2N$ alleles in the parent generation has an equal chance of having a copy appear in the offspring generation.

Problem 2.1 *What is the probability that a particular allele gets a copy into the next generation? The probability is one minus the probability that it doesn't make it. The surprising answer quickly becomes independent of the population size as N increases. (Use $\lim_{m\to\infty}(1 - 1/m)^m = e^{-1}$ to remove the dependence on population size.)*

You may have noticed that the computer simulations do not explicitly incorporate either segregation or demographic stochasticity, even though these two sources of randomness are the causes of genetic drift. Nonetheless, they do represent genetic drift as conceived by most population geneticists. A more realistic simulation with both sources of randomness would behave almost exactly like our simple one. Why then, do we use the simpler simulation? The answer is a recurring one in population genetics: The simpler model is easier to

* *Stochastic* is a synonym for *random*.

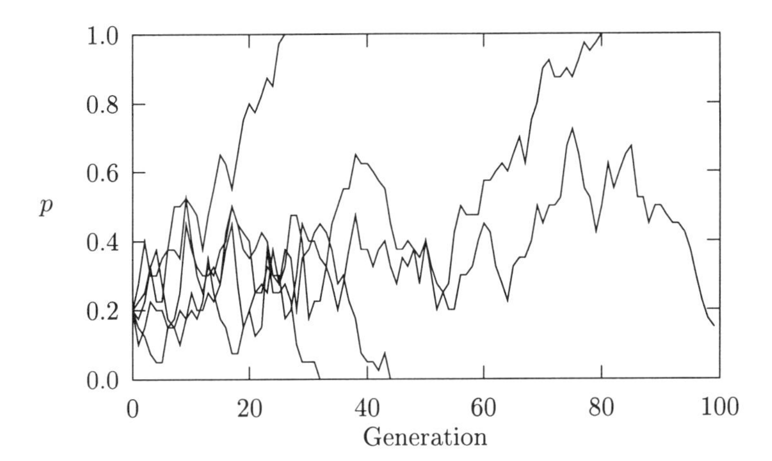

Figure 2.1: A computer simulation of genetic drift. The frequency of the A_1 allele, p, is graphed for 100 generations in five replicate populations each of size $N = 20$ and with initial allele frequency $p = 0.2$.

understand, is easier to analyze mathematically, and captures the essence of the biological situation. With drift, the essence is that each allele in the parental generation is equally likely to appear in the offspring generation. In addition, the probability that a particular allele appears in the offspring generation is nearly independent of the identities of other alleles in the offspring generation. The simple algorithm in the simulation does have both of these properties and simulates what is called the Wright-Fisher model in honor of Sewall Wright and R. A. Fisher, two pioneers of population genetics who were among the first to investigate genetic drift.

Important features of genetic drift are illustrated in Figure 2.1. One, of course, is that genetic drift causes random changes in allele frequencies. Each of the five populations behaves differently even though they all have the same initial allele frequency and the same population size. By implication, evolution can never be repeated. A second feature is that alleles are lost from the population. In two cases, the A_1 allele was lost; in two other cases, the A_2 allele was lost. In the fifth case, both alleles are still in the population after 100 generations. From this we might reasonably conclude that genetic drift removes genetic variation from populations. The third feature is more subtle: the direction of the random changes is neutral. There is no systematic tendency for the frequency of alleles to move up or down. A few simulations cannot establish this feature with certainty. That must wait for our mathematical development, beginning in the next section.

Problem 2.2 *If you know how to program a computer, write a simulation of genetic drift.*

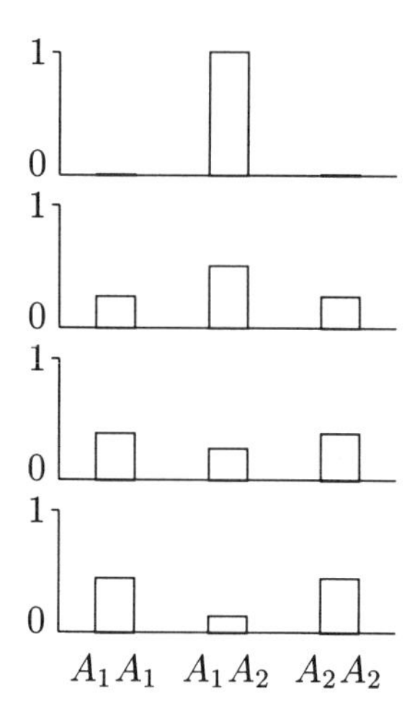

Figure 2.2: Genotype frequencies for four generations of drift with $N = 1$.

2.2 The decay of heterozygosity

As a warm-up to our general treatment of genetic drift, we will first examine the simplest non-trivial example of genetic drift: a population made up of a single hermaphroditic individual. If the individual is an A_1A_2 heterozygote, the frequency of the A_1 allele in the population is one-half. When the population reproduces by mating at random (a strange notion, but accurate) and the size of the population is kept constant at one, the heterozygote is replaced by an A_1A_1, A_1A_2, or A_2A_2 individual with probabilities 1/4, 1/2, and 1/4, respectively. These probabilities are the probabilities that the allele frequency becomes 1, 1/2, or 0 after a single round of random mating. In the first or third outcome, the population is a single homozygote individual and will remain homozygous forever. In the second outcome, the composition of the population remains unchanged.

After another round of random mating, the probability that the population is a heterozygous individual is 1/4, which is the probability that it is heterozygous in the second generation, 1/2, times the probability that it is heterozygous in the third generation given that it is heterozygous in the second generation, 1/2. The probabilities for the first four generations are illustrated in Figure 2.2. It should be clear from the figure that the probability that the population is a heterozygote after t generations of random mating is $(1/2)^t$, which approaches zero as t increases. On average, it takes only two generations for the population to become homozygous. When it does, it is as likely to be homozygous for the A_1 allele as for the A_2 allele.

While simple, this example suggests the following features about genetic drift, some of which overlap our observations on Figure 2.1.

- Genetic drift is a random process. The outcome of genetic drift cannot be stated with certainty. Rather, either probabilities must be assigned to different outcomes or the average outcome must be described.

- Genetic drift removes genetic variation from the population. The probability that an individual chosen at random from the population is heterozygous after t generations of random mating is

$$\mathcal{H}_t = \mathcal{H}_0 \left(1 - \frac{1}{2N}\right)^t,$$

where $\mathcal{H}_0$ is the initial probability of being a heterozygote (one in our example) and N is the population size (one in our example). This provocative form for our simple observation that $\mathcal{H}_t = (1/2)^t$ adumbrates the main result of this section.

- The probability that the ultimate frequency of the A_1 allele is one is equal to the frequency of the A_1 allele in the starting population, one-half.

Problem 2.3 *Convince yourself that the average time for the population to become homozygous is, in fact, two generations.*

The mathematical description of genetic drift can be quite complicated for populations with more than one individual. Fortunately, there is a simple and elegant way to study one of the most important aspects of genetic drift: the rate of decay of heterozygosity. As usual, we will be studying an autosomal locus in a randomly mating population made up of N diploid hermaphroditic individuals. The state of the population will be described by the variable $\mathcal{G}$, defined to be the probability that two alleles different by origin (equivalently, drawn at random from the population without replacement) are identical by state. These alleles are assumed to be completely equivalent in function and, thus, equally fit in the eyes of natural selection. Such alleles are called neutral alleles. $\mathcal{G}$ is a measure of the genetic variation in the population, which is almost the same as the homozygosity of the population as defined in Equation 1.2. When there is no genetic variation, $\mathcal{G} = 1$. When every allele is different by state from every other allele, $\mathcal{G} = 0$.

The value of $\mathcal{G}$ after one round of random mating, $\mathcal{G}'$, as a function of its current value, is

$$\mathcal{G}' = \frac{1}{2N} + \left(1 - \frac{1}{2N}\right)\mathcal{G}. \tag{2.1}$$

The derivation, illustrated in Figure 2.3, goes as follows. $\mathcal{G}'$ is the probability that two alleles that are different by origin in the next generation, called generation $t+1$ in Figure 2.3, are identical by state. Identity by state could happen in two different ways. One way is when the two alleles are copies of the same allele in the previous generation, as illustrated by the left-hand side of Figure 2.3. The probability that the two alleles do share an ancestor allele in the previous generation is $1/(2N)$. (Pick one allele, and the probability that

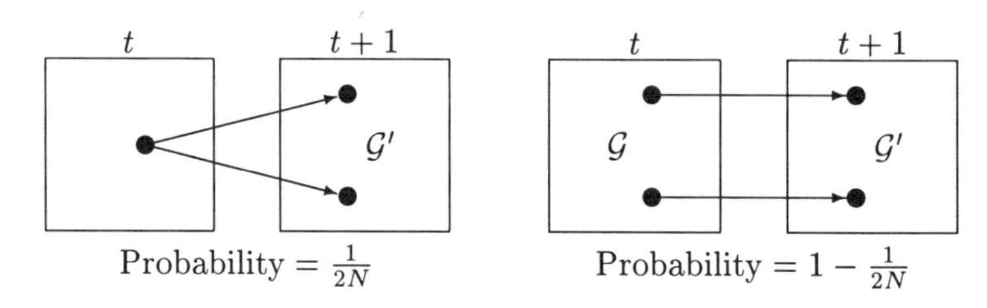

Figure 2.3: The derivation of $\mathcal{G}'$. The circles represent alleles; the arrows indicate the ancestry of the alleles.

the allele picked next has the same parent allele as the first is $1/(2N)$, as all alleles are equally likely to be chosen.) The second way for the alleles to be identical by state is when the two alleles do not have the same ancestor allele in the previous generation, as illustrated by the right-hand side of Figure 2.3, but their two ancestor alleles are themselves identical by state. This ancestry occurs with probability $1 - 1/(2N)$, and the probability that the two ancestor alleles are identical by state is $\mathcal{G}$ (by definition). As these two events are independent, the probability of the second way is $[1 - 1/(2N)]\mathcal{G}$. Finally, as the two ways to be identical by state are mutually exclusive, the full probability of identity by state in the next generation is obtained by summation, as seen in the right-hand side of Equation 2.1.

The time course for $\mathcal{G}$ is most easily studied by using

$$\mathcal{H} = 1 - \mathcal{G},$$

the probability that two randomly drawn alleles are different by state. ($\mathcal{H}$ is similar to the heterozygosity of the population.) From Equation 2.1 and a few algebraic manipulations, we have

$$\mathcal{H}' = 1 - \mathcal{G}' = \left(1 - \frac{1}{2N}\right)\mathcal{H},$$

and finally,

$$\boxed{\Delta_N \mathcal{H} = -\frac{1}{2N}\mathcal{H}} \tag{2.2}$$

The Δ operator is used to indicate the change in a state variable that occurs in a single generation, $\Delta_N \mathcal{H} = \mathcal{H}' - \mathcal{H}$. The subscript N in $\Delta_N \mathcal{H}$ is simply a reminder that the change is due to genetic drift.

Equation 2.2 shows that the probability that two alleles are different by state decreases at a rate $1/(2N)$ each generation. For very large populations, this decrease will be very slow. Nonetheless, the eventual result is that all of the variation is driven from the population by genetic drift.

The full time course for $\mathcal{H}$ is

$$\mathcal{H}_t = \mathcal{H}_0 \left(1 - \frac{1}{2N}\right)^t, \tag{2.3}$$

where $\mathcal{H}_t$ is $\mathcal{H}$ in the tth generation. The easiest way to show this is to examine the first few generations,

$$\begin{aligned} \mathcal{H}_1 &= \mathcal{H}_0 \left(1 - \frac{1}{2N}\right) \\ \mathcal{H}_2 &= \mathcal{H}_1 \left(1 - \frac{1}{2N}\right) \\ &= \mathcal{H}_0 \left(1 - \frac{1}{2N}\right)^2, \end{aligned}$$

and then make a modest inductive leap to the final result.

Equation 2.3 shows that the decay of $\mathcal{H}$ is geometric. The probability that two alleles are different by state goes steadily down but does not hit zero in a finite number of generations. Nonetheless, the probability eventually becomes so small that most populations will, in fact, be homozygous. Every allele will be a descendent of a single allele in the founding population. All but one of the possibly thousands or millions of alleles in any particular population will fail to leave any descendents.

Problem 2.4 *Graph $\mathcal{H}_t$ and $\mathcal{G}_t$ for 100 generations using $N = 1$, $N = 10$, $N = 100$, and $N =$ 1,000,000.*

For large populations, genetic drift is a very weak evolutionary force, as may be shown by the number of generations required to reduce $\mathcal{H}$ by one-half. This number is the value of t that satisfies the equation $\mathcal{H}_t = \mathcal{H}_0/2$,

$$\frac{\mathcal{H}_0}{2} = \mathcal{H}_0 \left(1 - \frac{1}{2N}\right)^t.$$

Cancel $\mathcal{H}_0$ from both sides, take the natural logarithm of both sides, and solve for t to obtain

$$t_{1/2} = \frac{-\ln(2)}{\ln(1 - 1/2N)}. \tag{2.4}$$

The approximation of the log given in Equation A.3,

$$\ln(1 + x) \approx x,$$

allows us to write

$$t_{1/2} \approx 2N \ln(2). \tag{2.5}$$

In words, the time required for genetic drift to reduce $\mathcal{H}$ by one-half is proportional to the population size.

When studying population genetics, placing results in a more general context is often enlightening. For example, a population of one million individuals requires about 1.38×10^6 generations to reduce $\mathcal{H}$ by one-half. If the generation time of the species were 20 years, it would take about 28 million years to halve the genetic variation. In geologic terms, 28 million years ago Earth was in the Oligocene epoch, the Alps and the Himalayas were rising from the collision of India and Eurasia and large browsing mammals first appeared, along with the first monkey-like primates. During the succeeding 28 million years, whales, apes, large carnivores, and hominoids all appeared, while genetic drift was poking along removing one-half of the genetic variation.

Problem 2.5 *Graph simultaneously both Formula 2.4 and Formula 2.5 as a function of N for N from 1 to 100. Is the approximation to your liking?*

Another property of genetic drift that is easy to derive is the probability that the A_1 allele will be the sole surviving allele in the population. This probability is called the fixation probability. In Figure 2.1, the A_1 allele was fixed in two of the four replicate populations in which a fixation occurred. We could use the simulation to guess that the fixation probability of the A_1 allele is about one-half. In fact, the fixation probability is 0.2, as will emerge from a few simple observations.

As all variation is ultimately lost, we know that eventually one allele will be the ancestor of all of the alleles in the population. As there are $2N$ alleles in the population, the chance that any particular one of them is the ancestor of all (once $H = 0$) is just $1/(2N)$. If there were, say, i copies of the A_1 allele, then the chance that one of the i copies is the ancestor is $i/(2N)$. Equivalently, if the frequency of the A_1 allele is p, then the probability that all alleles are ultimately A_1 is p. In this case, we say that the A_1 allele is fixed in the population. Thus, the probability of ultimate fixation of a neutral allele is its current frequency, $\pi(p) = p$, to introduce a notation that will be used later in the book. This is as trivial as it is because all alleles are equivalent; there is no natural selection.

Notice that our study of $\mathcal{H}$ and the fixation probability agrees very well with the observations made earlier on the population composed of a single individual. You should use the simple case whenever your intuition for the more complicated case needs help.

Problem 2.6 *$\mathcal{G}$ is almost the same as the homozygosity of the population, G. Suppose we were to define the homozygosity of a population as the probability that two alleles chosen at random from the population with replacement are identical by state. Show that this is equivalent to the definition given in Equation 1.2. Next, show that*

$$G = \frac{1}{2N} + \left(1 - \frac{1}{2N}\right) \mathcal{G}.$$

Use this to justify the claim that G and $\mathcal{G}$ are "almost the same." It should be clear that we could have used the term heterozygosity *everywhere that we used $\mathcal{H}$ without being seriously misled.*

Genetic drift is an evolutionary force that changes both allele and genotype frequencies. No population can escape its influence. Yet it is a very weak evolutionary force in large populations, prompting a great deal of debate over its relative importance in evolution. Drift is undeniably important for the dynamics of rare alleles, in small subdivided populations with very low migration rates, and in the neutral theory of molecular evolution. Beyond these three arenas, there is little agreement about the importance of drift.

Genetic drift appears to call into question the validity, or at least the utility, of the Hardy-Weinberg law. However, this is not the case except in the smallest of populations. The attainment of Hardy-Weinberg frequencies takes only a generation or two. Viewed as an evolutionary force, random mating has a time scale of one or two generations. Drift has a time scale of $2N$ generations, vastly larger than one or two for natural populations. When two forces have such different time scales, they rarely interact in an interesting way. This is certainly true for the interaction of drift and random mating. In any particular generation, the population will appear to be in Hardy-Weinberg equilibrium. The deviation of the frequency of a genotype from the Hardy-Weinberg expectation will be no more than about $1/(2N)$, certainly not a measurable deviation. Moreover, the allele frequency will not change by a measurable amount in a single generation. Thus, there is nothing that an experimenter could do to tell, based on allele and genotype frequencies, that the population does not adhere faithfully to the Hardy-Weinberg predictions.

2.3 Mutation and drift

If genetic drift removes variation from natural populations, why aren't all populations devoid of genetic variation? The answer, of course, is that mutation restores the genetic variation that genetic drift eliminates. The interaction between drift and mutation is particularly important for molecular population genetics and its neutral theory. The neutral theory claims that most of the DNA sequence differences between alleles within a population or between species are due to neutral mutations. The mathematical aspects of the theory will be developed in this section. The following section will bring the mathematics and the data together.

Mutation introduces variation into the population at a rate $2Nu$, where u is the mutation rate to neutral alleles. Genetic drift gets rid of variation at a rate $1/(2N)$. At equilibrium, the probability that two alleles different by origin are identical by state is given by the classic formula

$$\boxed{\hat{\mathcal{G}} = \frac{1}{1+4Nu}} \tag{2.6}$$

There are many ways to obtain Formula 2.6. We will use a traditional approach that follows directly from Equation 2.1 with the addition of mutation.

The probability that a mutation appears in a gamete at the locus under study is u, which is called the mutation rate even though mutation probability would be a more accurate term. When a mutation does occur, it is assumed to be to a unique allele, one that differs by state from all alleles that have ever existed in the population. A population that has been around for a long time will have seen a very large number of different alleles. Consequently, our model of mutation is often called the infinite-allele model. The infinite-allele model is meant to approximate the large, though finite, number of alleles that are possible at the molecular level.

Problem 2.7 *How many different alleles are one mutational step away from an allele at a locus that is 3000 nucleotides in extent? How many are two mutational steps away?*

The value of $\mathcal{G}$ after one round of random mating and mutation, as a function of its current value, is

$$\mathcal{G}' = (1-u)^2 \left[\frac{1}{2N} + \left(1 - \frac{1}{2N}\right)\mathcal{G}\right]. \tag{2.7}$$

Notice that this formula differs from Equation 2.1 only by the factor $(1-u)^2$ on the right side. This factor is the probability that a mutation did not occur in either of the two alleles chosen at random from the population in the "prime" generation. (The probability that a mutation did not occur in one allele is $1-u$; the two alleles are independent.)

Equation 2.7 may be manipulated to obtain an approximation for $\Delta\mathcal{H}$. The approximation is based on the fact that mutation rates are small (often 10^{-5} to 10^{-10} depending on the context) and population sizes are large (often much larger than 10^4). The approximation is obtained in three steps. For the first two, approximate $(1-u)^2$ with $1-2u$ and ignore terms with u/N as factors,

$$\begin{aligned}\mathcal{G}' &\approx (1-2u)\left[\frac{1}{2N} + \left(1 - \frac{1}{2N}\right)\mathcal{G}\right] \\ &\approx \frac{1}{2N} + \left(1 - \frac{1}{2N}\right)\mathcal{G} - 2u\mathcal{G}.\end{aligned}$$

For the third step, set $\mathcal{H} = 1 - \mathcal{G}$ and rearrange a bit to get

$$\mathcal{H}' \approx \left(1 - \frac{1}{2N}\right)\mathcal{H} + 2u(1-\mathcal{H}).$$

The change in $\mathcal{H}$ in a single generation is

$$\Delta\mathcal{H} \approx -\frac{1}{2N}\mathcal{H} + 2u(1-\mathcal{H}). \tag{2.8}$$

At equilibrium, $\Delta\mathcal{H} = 0$. The value of $\mathcal{H}$ that satisfies $\Delta\mathcal{H} = 0$ is

$$\hat{\mathcal{H}} = \frac{4Nu}{1+4Nu}, \tag{2.9}$$

which gives Equation 2.6 immediately.

Problem 2.8 *Graph Equation 2.6 as a function of $4Nu$. What value of $4Nu$ gives a reasonable fit to the average heterozygosity for proteins as described by electrophoretic studies in humans?*

Our route to the equilibrium value of $\mathcal{H}$ was rapid, as befits such a simple result. Now, we must return to the derivation to bring out a few features that will increase our understanding of the interaction of evolution forces. Examination of Equation 2.8 shows that the change in $\mathcal{H}$ is a sum of two components, which may be written as

$$\Delta\mathcal{H} \approx \Delta_N\mathcal{H} + \Delta_u\mathcal{H}.$$

The left component,

$$\Delta_N\mathcal{H} = -\frac{1}{2N}\mathcal{H},$$

is the already familiar change in $\mathcal{H}$ due to genetic drift. The right component,

$$\Delta_u\mathcal{H} = 2u(1 - \mathcal{H}), \tag{2.10}$$

is the change in $\mathcal{H}$ due to mutation and may be derived by considering the effects of mutation in isolation, that is, in an infinite population.

In an infinite population with mutation, the probability that two randomly chosen alleles are different in state in the next generation is

$$\mathcal{H}' = \mathcal{H} + (1 - \mathcal{H})[1 - (1 - u)^2],$$

which is the sum of the two ways that the two alleles may have come to be different. The first way is if their parent alleles were different by state, which occurs with probability $\mathcal{H}$. The second way is if their parent alleles were not different and a mutation occurred in the production of at least one of the two alleles. The mutation term on the far right is one minus the probability that no mutation occurred, which is the same as the probability that at least one mutation occurred. By approximating $(1-u)^2$ with $1-2u$ and performing some minor rearrangements, we recover Equation 2.10 for $\Delta_u\mathcal{H}$.

The change due to drift is always negative (drift decreases genetic variation), whereas the change due to mutation is always positive (mutation increases genetic variation). Equilibrium is reached when these two evolutionary forces exactly balance, $-\Delta_N\mathcal{H} = \Delta_u\mathcal{H}$. The equilibrium is interesting only when $4Nu$ is moderate in magnitude. If $4Nu$ is very small, then genetic drift dominates mutation and genetic variation is eliminated from the population. If $4Nu$ is very large, mutation dominates drift and all $2N$ alleles in the population are unique in state.

The comparison of the strengths of evolutionary forces is key to an understanding of their interaction. The strength of a force may be quantified by the time required for the force to have a significant effect on the genetic structure of the population. In the previous section, we saw that the time required for genetic drift to reduce the heterozygosity of the population by one-half is proportional to the population size, N. We could say that the time scale of genetic drift is proportional to the population size. Similarly, the time scale of mutation is proportional to $1/u$. If $1/u \ll N$, the time scale of mutation is much less than that of drift, leading to a population with many unique alleles. If $N \ll 1/u$, the time scale of drift is shorter, leading to a population devoid of variation.

Problem 2.9 *Graph $-\Delta_N \mathcal{H}$ and $\Delta_u \mathcal{H}$ as function of $\mathcal{H}$ for $N = 10^4$ and $u = 5 \times 10^{-5}$. Do the lines intersect where you expect them to?*

In addition to finding the equilibrium properties of the population, we can also derive the rate of substitution of neutral mutations. The rate of substitution could just as well have been called the rate of fixation, as it is a measure of how frequently genetic drift and mutation cause alleles to become fixed in the population. To calculate the rate of substitution, we need only multiply the average number of mutations that enter the population each generation by the fraction of those mutations that fix.

The average number of new mutations entering the population each generation is equal to the number of gametes produced each generation times the probability of a mutation in any one of them, $2Nu$. Thus, in each generation there will be, on average, $2Nu$ new mutations in the population. The simplicity of this result often obscures an important consequence: more mutations enter large populations each generation than enter small populations. One might expect evolution to proceed more rapidly in large than small populations. This is often the case, but not when dealing with neutral evolution, as we are in this chapter.

Of the new mutations that enter the population each generation, a fraction, $1/(2N)$, will fix, on average. This follows from our previous observation that the chance that any particular allele will fix in a population is equal to its frequency, which is $1/(2N)$ for a new mutation. Thus, the average rate of substitution, k, is $2Nu \times 1/(2N)$ or

$$\boxed{k = u} \tag{2.11}$$

In words, the rate of substitution of neutral alleles, k, is equal to the mutation rate to neutral alleles, u. This is one of the most remarkable results in all of population genetics. At first it offends our intuition. The fixation of alleles is caused by genetic drift. The strength of genetic drift depends on the population size. Our intuition wants the rate of fixation to depend on population size as well. However, the number of mutations entering the population each generation also depends on population size and does so in such a way as to cancel out drift's dependency on population size when deriving the rate of substitution.

There are a few subtleties in the interpretation of the neutral mutation rate, $k = u$. In a population that is so large that $4Nu \gg 0$, a new mutant allele will almost never fix in the population simply because a large number of new mutations enters the population each generation. If there are no fixations, how can Equation 2.11 be interpreted as the rate of substitution? To understand this paradox, we need an explicit model of a locus. One model in common use in

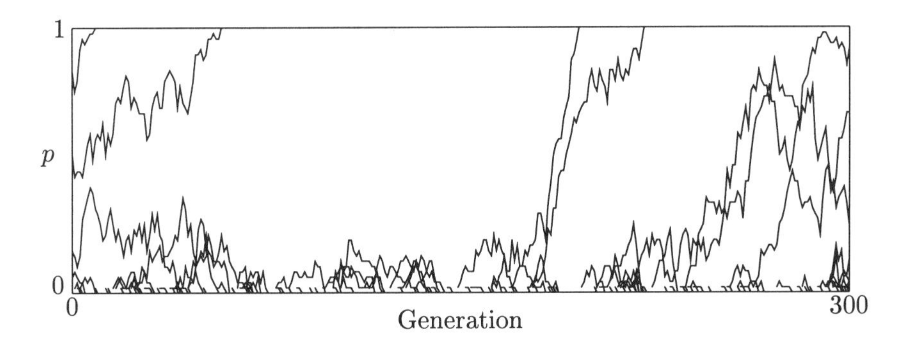

Figure 2.4: A computer simulation of the infinite-sites, free-recombination model of a locus experiencing neutral evolution in a population of 25 individuals with $4Nu = 0.5$.

population genetics, the infinite-sites, free-recombination model, imagines that a locus is a very large (effectively infinite) sequence of nucleotides, each evolving independently. Independent evolution requires that there is a fair amount of recombination between nucleotides, hence the "free recombination" part of the name. The most important consequence of having a large number of nucleotides is that the mutation rate per nucleotide is very small. For example, if the mutation rate at a locus of 1000 nucleotides were 10^{-6}, the mutation rate at each nucleotide would be 10^{-9}. With such a small mutational input per site, it is plausible that mutations at sites will fix, even if there is never a fixation of an allele at the locus. If we call the mutation rate per site u_s, then the rate of substitution per site is $k_s = u_s$. If there are n sites in the locus, then the rate of substitution for the locus is

$$k = nk_s = nu_s = u,$$

where u is the locus mutation rate.

Figure 2.4 presents the output of a computer simulation of the infinite-sites, free-recombination model. The vertical axis is the frequency of mutations at sites. During the course of the simulation, there were fixations at four sites. However, at no time was the population homozygous, so there could not have been any fixations of alleles.

The infinite-sites model may be immediately applied to DNA sequence data. For example, in the previous chapter we saw that *Drosophila melanogaster* and *D. erecta* differed by 36 of 768 nucleotides. If we assume that the two species separated 23 million years ago and that there were exactly 36 substitutions on the lineages leading to the two species,* then the rate of substitution per site is

$$\begin{aligned} k_s &= \frac{36/768 \text{ substitutions/site}}{2 \times 23{,}000{,}000 \text{ years}} \\ &\approx 10^{-9} \text{ substitutions/site/year.} \end{aligned}$$

*There may have been more than 36 substitutions as multiple substitutions at a site are not detectable.

(The factor of two in the denominator is because the total time separating the two species is the sum of the lengths of the branching leading from their common ancestor to each species.) If we assume that the substitutions are neutral, then the neutral mutation rate must be $u_s = 10^{-9}$ mutations per site per year. If *Drosophila* has one generation per year, then the mutation rate per generation is also 10^{-9}. If, on the other hand, *Drosophila* has two generations per year, then the mutation rate per generation must be $(1/2) \times 10^{-9}$ so that the rate of neutral substitution matches the rate of substitution per year estimated from the sequence data.

Because the neutral rate of substitution depends only on the mutation rate, it is natural to expect that the rate of substitution will be approximately the same for different groups of species. In fact, the apparent constancy of evolutionary rates of proteins has long been used to argue that most evolution at the molecular level is due to drift and mutation alone. The neutral theory is concerned with this hypothesis.

2.4 The neutral theory

The neutral theory of molecular evolution claims that most allelic variation and substitutions in proteins and DNA are neutral. Neutral evolution has been called non-Darwinian evolution because most substitutions are due to genetic drift rather than natural selection. However, the theory is not in conflict with Darwin's theory; rather, it simply claims that most substitutions have no influence on the survival of genotypes. Those few that do are subject to natural selection and change the adaptation of the species to its environment. The neutral theory was controversial when first proposed and remains controversial today. Many individuals proposed the theory, more or less independently, in the 1960s. Among these were N. Sueoka, E. Freese, A. Robertson, M. Kimura, T. Jukes, and J. L. King. The idea is so simple that many others undoubtedly thought of it at about the same time. The first paper to develop fully the population genetics aspects of the theory was the 1971 paper by Motoo Kimura and Tomoko Ohta, "Protein polymorphism as a phase of molecular evolution." This paper is a true classic in population genetics and will be examined in detail in this section.

The first section of the paper, "Rate of Evolution," points out that the rate of amino acid substitution per year is remarkably constant among vertebrate lineages for each protein examined. This constancy led to the concept of a molecular clock which, like the ticking of a very slow clock, is the roughly constant rate of occurrence of substitutions through geologic time. Hemoglobins, as illustrated in Figure 2.5, experience about one amino acid substitution every one billion years at each amino acid site in vertebrate lineages. Thus, the rate of substitution in hemoglobins is approximately $k_s = 10^{-9}$ amino acid substitutions per amino acid site per year.* Cytochrome c evolves more slowly,

*In this section, we will use the subscript s on substitution and mutation rates when they refer to sites. Kimura and Ohta do not use subscripts.

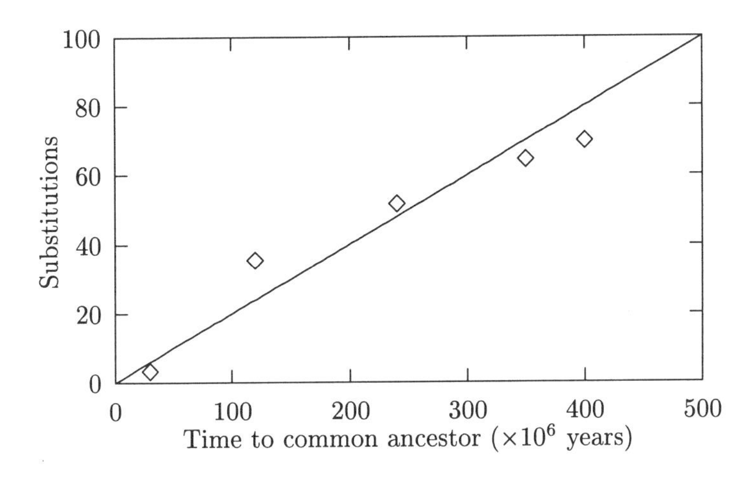

Figure 2.5: The number of amino acid substitutions in beta globin that occurred in the lineages leading to humans and various species as a function of the time back to their common ancestors.

about 0.3 substitution per site per billion years, or $k_s = 0.3 \times 10^{-9}$. The average rate among the proteins examined was $k_s = 1.6 \times 10^{-9}$. The apparent constancy of the rate of substitution is viewed by Kimura and Ohta as strong evidence in favor of the neutral theory. They argue that natural selection, being more opportunistic and more subject to the vagaries of the environment, might be expected to cause a much more erratic pattern of substitutions.

If substitutions are neutral, we know from Equation 2.11 that the average neutral mutation rate, u_s, must be 1.6×10^{-9} amino acid mutations per amino acid site per year, which is strikingly close to nucleotide mutation rates as measured in the laboratory. Is this coincidence, or is amino acid evolution neutral with $k_s = u_s \approx 10^{-9}$? Kimura and Ohta show how polymorphism data may be used to support the hypothesis that most of the protein evolution is, in fact, neutral.

The next two sections of the paper, "Polymorphism in Sub-populations" and "Mutation and Mobility" deal with technical points and may be skipped on first reading. In the following section, "Heterozygosity and Probability of Polymorphism," Kimura and Ohta argue that, as typical protein heterozygosities are around 0.1, $4Nu$ must be approximately 0.1 (see their equation 8). You should be able to verify this using our Equation 2.9.

The next section, "Relative Neutral Mutation Rate," is the most important as it puts the substitution and polymorphism estimates together. The first issue concerns the fact that two mutation rates are used; u in the expression $4Nu \approx 0.1$ refers to the electrophoretically detectable mutations in the entire protein, while u_s in the expression $k_s = u_s = 1.6 \times 10^{-9}$ refers to all of the amino acid mutations at a typical site in the protein. As electrophoresis does not detect all of the variation that is present, some adjustment must be made so that the mutation rates are equivalent. Kimura and Ohta argue that a typical protein is about 300 amino acids long and that about 30 percent of the variation is detected by electrophoresis. Thus, the mutation rate to electrophoretically detectable variation for the entire protein is

$$u = 1.6 \times 10^{-9} \times 300 \times 0.3 = 1.44 \times 10^{-7}.$$

(Actually, their paper has an unfortunate typographical error that reports that $u = 1.6 \times 10^{-7}$.) The correct result, as given here, is viewed as an overestimate, so Kimura and Ohta use $u = 10^{-7}$ for the remainder of their investigation.

The electrophoretic mutation rate is in units of detectable mutations per protein per year, whereas Equation 2.9 for the equilibrium value of $\mathcal{H}$ requires the mutation rate per generation. The conversion is simple. For example, if mice have two generations per year, then the mutation rate per generation will be half of 10^{-7}. For mice, we have

$$4Nu = 4N \times 10^{-7}/2 \approx 0.1,$$

which gives $N \approx 5 \times 10^5$. The population size being estimated is called the effective population size, a concept that will be discussed in the next section. There was nothing in 1971 to suggest that the effective size of the house mouse was incompatible with about 10^5 individuals, despite our subjective impression that house mice exist in vast numbers throughout the world. Thus, the data on protein polymorphism and substitutions were viewed to be mutually compatible and to support the neutral theory.

Problem 2.10 *If the human generation time is 20 years, what effective population size for humans is implied by the data?*

Much has been written on the neutral theory since 1971. The theory slowly gained momentum through the 1970s, culminating in Kimura's book published in 1983. At that time, the neutral theory was the dominant explanation for most of protein and DNA evolution. More recently, the theory has fallen on hard times, particularly with regard to protein evolution. There are two aspects of the Kimura and Ohta paper that presented problems for the theory in later years. The first was the observation that the rate of substitution of amino acids is roughly constant per year. Our derivation of $k = u$ was done in time units of generations, not years. Thus, the rate of substitution should be roughly constant per generation rather than per year. As a consequence, creatures with shorter generation times should evolve faster than those with longer generation times. This generation-time effect, a clear prediction of the neutral theory, was not observed in proteins but was seen in noncoding DNA. This point, glossed over in the Kimura and Ohta paper, later caused a major shift in the neutral theory. The new version, due mainly to Tomoko Ohta, assumes that most amino acid substitutions are not neutral but are slightly deleterious. We will return to this theory in the next chapter.

The second point concerns the estimates of the effective population sizes using heterozygosity estimates. When the paper was written, most species that had been examined, from *Homo* to *Drosophila*, had heterozygosities in the narrow range of $0.05 < H < 0.15$. Figure 1.4 shows this to be true for the averages across species in most major taxonomic groups. By implication, the effective population sizes of these species are also in a narrow range, in contradiction to the common belief that the population sizes of different species vary over several orders of magnitude. Compounding this problem is the fact that the effective sizes of some species, like *Drosophila*, might well be greater than 10^{10}. If the neutral theory were true, then the heterozygosity of *Drosophila* should be much higher than it is.

There are two refinements of the neutral theory that can explain the narrow range of heterozygosities. The first is Ohta's hypothesis that amino acid mutations are slightly deleterious and thus are less frequent than predicted under the strictly neutral model. Her theory will be explained in the next chapter. The second involves the concept of the effective population size as developed in the next section.

2.5 Effective population size

The model captured in Equation 2.1 contains explicit and implicit assumptions about the population that appear to compromise its usefulness. For example, the species is assumed to be hermaphroditic and randomly mating. A more egregious assumption is the constancy of the population size. Not only do real populations fluctuate in size, they often fluctuate wildly. Do these simplifying assumptions render our investigations irrelevant, or can we make some minor adjustments and continue with our hard-won insights about genetic drift? In many cases, we can make some adjustments by using the concept of the effective population size.

In most models of natural populations, no matter how complex, the heterozygosity eventually decreases geometrically, just as it does in our idealized population. However, the rate of decrease will no longer be $1/(2N)$, but will be some new rate, call it $1/(2N_e)$, that depends on the particulars of the model. The parameter N_e is called the effective size of the population. It is the size of an idealized population whose rate of decay of heterozygosity is the same as that of the complicated population. Thus, we need only investigate how each complicating assumption influences the effective size of the population. From then on, we can simply substitute N_e for N in all of the preceding equations.

Of all the factors that affect the effective size, none is more important than fluctuations in the actual population size. Suppose, for example, the population sizes form a sequence $N_1, N_2, \ldots$ indexed by the generation number. The value of $\mathcal{H}$ in generation $t+1$, as a function of the population size in the previous generation, is

$$\mathcal{H}_{t+1} = \mathcal{H}_t \left(1 - \frac{1}{2N_t}\right).$$

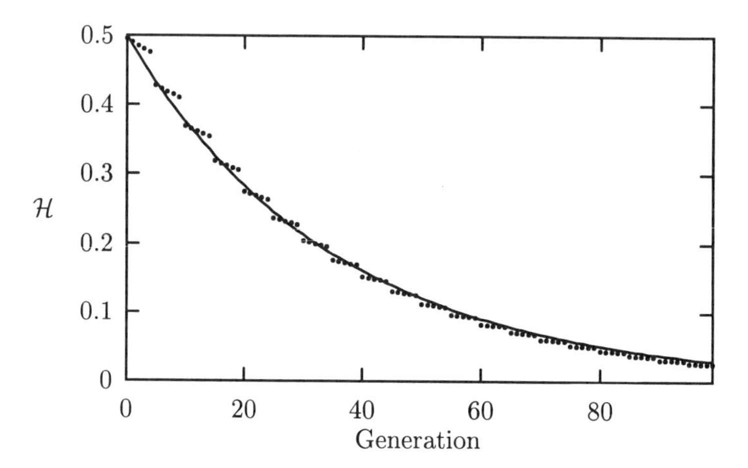

Figure 2.6: The time course for $\mathcal{H}$ in two populations. The dots are from a population with changing size with a harmonic mean of 17.875. The lines are from a population whose size is constant at 17.875 individuals.

Using the same argument that resulted in Equation 2.3, we obtain

$$\mathcal{H}_t = \mathcal{H}_0 \prod_{i=0}^{t-1} \left(1 - \frac{1}{2N_i}\right).$$

At this point, we can use the definition of the effective population size to argue that the effective size of a population with fluctuating actual population sizes satisfies the equation

$$\mathcal{H}_0 \left(1 - \frac{1}{2N_e}\right)^t = \mathcal{H}_0 \prod_{i=0}^{t-1} \left(1 - \frac{1}{2N_i}\right). \tag{2.12}$$

On the left side is $\mathcal{H}$ of an idealized population of actual size N_e and on the right side is $\mathcal{H}$ for the population complicated by population size fluctuations. The effective size of the complicated population is found by solving Equation 2.12 for N_e.

Equation 2.12 may be solved by using the techniques of Appendix A for approximating the product of terms that are close to one. The left side of Equation 2.12 is approximately $\exp(-t/2N_e)$, which is derived with the help of Equation A.3. The right-hand side is covered by Equation A.4. Together we get

$$e^{-t/2N_e} \approx e^{-\sum_{i=0}^{t-1} 1/2N_i}.$$

Equating the exponents and solving for the effective size, we have,

$$N_e \approx \left(\frac{1}{t} \sum_{i=0}^{t-1} \frac{1}{N_i}\right)^{-1}, \tag{2.13}$$

which is our final expression for the effective size of the population. You might recognize the formula for N_e as being the harmonic mean population size. The harmonic mean of the numbers $x_1, x_2, \ldots, x_n$ is the reciprocal of the average of the reciprocals of the x_i,

$$\frac{1}{\frac{1}{n}\left(\frac{1}{x_1}+\frac{1}{x_2}+\cdots+\frac{1}{x_n}\right)}.$$

One of the classic inequalities of mathematics states that the harmonic mean of a sequence of numbers is always less than or equal to the arithmetic mean.

Figure 2.6 illustrates the decay of $\mathcal{H}$ in a population whose size oscillates such that it is at 50 individuals for four generations, then 5 individuals for one generation, then 50 for four, and so forth. The figure also shows the decay of variation in a population of constant size 17.875, which is the harmonic mean population size of the oscillating population. The two populations clearly lose their variation at the same rate, even though they do not share exactly the same values for $\mathcal{H}$.

The result that the effective size equals the harmonic mean population size is quite important because the harmonic mean is much more sensitive to small values than is the arithmetic mean. For example, suppose the population size is 1000 for nine generations and 10 for one generation. The arithmetic mean in this case is

$$\frac{9}{10}\times 1000+\frac{1}{10}\times 10 = 901.$$

The harmonic mean is

$$\left(\frac{9}{10}\times\frac{1}{1000}+\frac{1}{10}\times\frac{1}{10}\right)^{-1} = 91.4,$$

which is about an order of magnitude smaller than the arithmetic mean population size. Were species to experience occasional but recurring crashes in population size, called bottlenecks by population geneticists, then the effective sizes of species would be considerably smaller than their "typical" sizes.

Population size fluctuations can account for two of the problems with the neutral theory mentioned in the previous section. The first is the problem of too little molecular variation based on our sense of current population sizes. If the sizes of most species do fluctuate wildly, then their effective sizes may be close to 10^5 as required by the neutral estimation, even though their current sizes are several orders of magnitude larger. The second problem concerns the narrow range of heterozygosities. Recall that the harmonic mean is determined more by the population size during bottlenecks than during times of abundance. However, there is a limit to how small population sizes can be, as rare species inevitably face extinction. In fact, extant species are precisely those species that have not suffered recurring catastrophic crashes in population sizes. The net effect is to make the harmonic means of species more similar than their actual numbers at any point in time. Although we should still be uncomfortable with the narrow range of heterozygosities, we can be less uncomfortable given insights provided by our analysis of the effective size of species.

Other factors also affect the effective size, but not so dramatically as bottlenecks. In randomly mating dioecious species with different numbers of males, N_m, and females, N_f, the effective size is

$$N_e = \frac{4N_m N_f}{N_f + N_m}.$$

For example, suppose the number of males is α times the number of females, $N_m = \alpha N_f$. The ratio of the effective size to the actual size is then

$$\frac{N_e}{N_m + N_f} = \frac{4\alpha}{(1+\alpha)^2}.$$

In a species with one-tenth the number of males as females, the effective size will be about 33 percent of the actual population size. This is not a particularly extreme reduction. Except for species with population sizes so small that extinction seems imminent anyway, a 33 percent reduction in the effective size is of little consequence. Our estimates of actual population sizes are so imprecise that the adjustment required for such factors as different numbers of the sexes often seem unimportant compared to the very large reductions that come with fluctuations in population size.

2.6 The coalescent

The parameter $\theta = 4Nu$ determines the level of variation under the neutral model. Assuming that the variation at a locus is neutral, θ may be estimated by using the observed homozygosity and Equation 2.6, as done, for example, by Kimura and Ohta. If the data come as DNA sequences rather than from protein electrophoresis, then it is possible to use the theory of coalescents and the additional information in the sequence to obtain a better estimate of θ.

A coalescent is the lineage of alleles in a sample traced backward in time to their common ancestor allele. The left side of Figure 2.7 illustrates a typical coalescent for a sample of four alleles. Each of the alleles in the sample is descended from an allele in the previous generation. In a large population, the immediate ancestors of the sampled alleles are likely to be distinct. However, if the ancestors of the sampled alleles are followed backward for a long time, eventually a generation will be found with a common ancestor of two of the alleles. In the figure, the two central alleles have a common ancestor t_1 generations in the past. At t_1 we say that a coalescence occurred because the number of lineages in the coalescent is reduced by one. The next coalescence occurs t_2 generations in the past, and the final coalescence occurs t_3 generations in the past. Time, when discussing coalescents, is always measured in units of generations in the past.

The coalescent is sometimes called the genealogy of the sample because it captures the genealogic relationships of the sampled alleles. The coalescent cannot be known, of course, because there is no way to know which alleles share

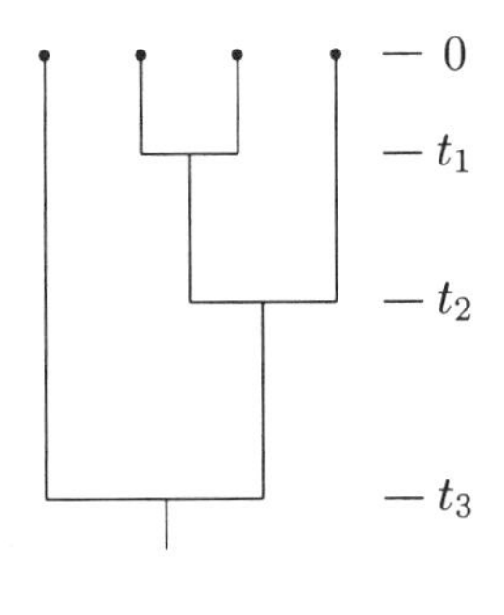

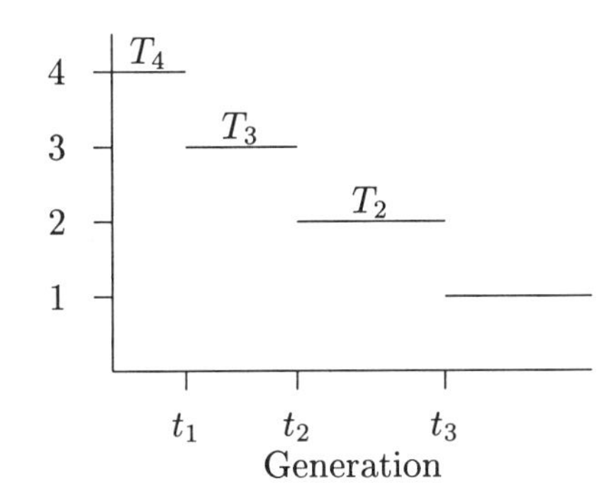

Figure 2.7: On the left is an example of a coalescent for four alleles. On the right is a graph showing the size of the coalescent as a function of time measured backward.

which common ancestors, nor is there any way to know the times of coalescence events. Coalescents are useful constructs because in natural populations mutations occur on the lineages and these mutations can be used to infer some of the coalescent and to estimate θ. Each mutation on the coalescent results in a segregating site in the sample. The number of segregating sites in a sample of n alleles, S_n, turns out to contain enough information to estimate θ.

Problem 2.11 *Convince yourself that each mutation results in a segregating site by drawing some coalescents, marking random spots where mutations occur, and then constructing the set of DNA sequences implied by those mutations.*

Although we cannot know t_1, t_2, and t_3 for a sample of four alleles, Figure 2.7 shows that it is possible to state that the total time in the coalescent, T_c, depends on these times through

$$\begin{aligned} T_c &= 4 \times t_1 + 3 \times (t_2 - t_1) + 2 \times (t_3 - t_2) \\ &= 4T_4 + 3T_3 + 2T_2, \end{aligned}$$

where T_i is the time required to reduce a coalescent with i alleles to one with $i-1$ alleles. The time course of the number of alleles in the coalescent is illustrated on the right side of Figure 2.7.

For a neutral mutation rate of u, the expected number of mutations in the coalescent is $T_c u$. As we will soon see, the expected value of T_c for a sample of four alleles is

$$E\{T_c\} = 4N\left(1 + \frac{1}{2} + \frac{1}{3}\right) = \frac{44N}{6},$$

which implies that, on average, there will be

$$uE\{T_c\} = \theta(11/6)$$

mutations in a sample of four alleles. Because each mutation in the coalescent contributes one segregating site, the expected number of segregating sites in a sample of four alleles is also

$$E\{S_4\} = \theta(11/6),$$

which suggests that $\hat{\theta} = (6/11)S_4$ should be a good estimator for θ in a sample of four alleles.*

With this motivation, it is now time to tackle the expected time in a coalescent for a sample of n alleles. The time interval for the first coalescence is called T_n, the time interval for the second coalescence, T_{n-1}, and so forth back to the time interval for the final coalescence, T_2. The mean length of each interval may be readily found if we know the probability that a coalescence does not occur in the previous generation or, as we prefer to call it, the next ancestral generation.

Consider the history of the n alleles in turn. The first allele has, of course, an ancestor allele in the first ancestral generation. The second allele will have a different ancestor allele with probability

$$1 - \frac{1}{2N} = \frac{2N-1}{2N}.$$

The right-hand side of this equation is particularly informative because it shows that there are a total of $2N$ possible ancestors for the second allele, but only $2N - 1$ that are different from the first allele. The probability that the third allele does not share an ancestor with the first two, assuming that the first two do not share an ancestor, is $(2N-2)/(2N)$; the total probability that the first three do not share an ancestor is

$$\frac{2N-1}{2N} \times \frac{2N-2}{2N}.$$

With an obvious leap of intuition, the probability that the n alleles all have different ancestors is

$$\left(1 - \frac{1}{2N}\right)\left(1 - \frac{2}{2N}\right)\cdots\left(1 - \frac{n-1}{2N}\right) \approx 1 - \frac{1}{2N} - \frac{2}{2N} - \cdots - \frac{n-1}{2N},$$

where terms of order N^{-2} and smaller have been ignored in the approximation. The probability that a coalescence occurs is one minus the probability that it does not, or

$$\frac{1 + 2 + \cdots + (n-1)}{2N} = \frac{n(n-1)}{4N},$$

where the final step uses the fact that the sum of the first m integers is

$$1 + 2 + \cdots + m = m(m-1)/2.$$

*A common convention in statistics notates the estimator of a parameter as the parameter with a hat over it.

If the probability that the first coalescence occurs in any particular generation is $n(n-1)/(4N)$, then the probability distribution of the time until the first coalescence is a geometric distribution with probability of success $p = n(n-1)/(4N)$. (See Appendix B for properties of the geometric distribution.) The mean of the geometric distribution is the reciprocal of the probability of success, giving

$$E\{T_n\} = \frac{4N}{n(n-1)}.$$

The beauty of the coalescent approach is that all of the hard work is now over. A little reflection shows that the mean time interval leading from a coalescent with i alleles to one with $i-1$ alleles is just

$$E\{T_i\} = \frac{4N}{i(i-1)}, \tag{2.14}$$

as there is nothing special about n in the previous derivation of ET_n.

The total time in all of the branches of a coalescent is

$$T_c = \sum_{i=2}^{n} iT_i,$$

which, using the fact that the expectation of the sum of random quantities is the sum of the expectations of those quantities (see Equation B.11 on page 162), is

$$E\{T_c\} = \sum_{i=2}^{n} iE\{T_i\} = 4N\sum_{i=2}^{n}\frac{1}{i-1}.$$

Recalling that the expected number of segregating sites is the neutral mutation rate, u, times the expected time in the coalescent, we have

$$E\{S_n\} = uE\{T_c\} = \theta\sum_{i=2}^{n}\frac{1}{(i-1)},$$

which suggests that

$$\hat{\theta} = \frac{S_n}{1 + \frac{1}{2} + \frac{1}{3}\cdots + \frac{1}{n-1}} \tag{2.15}$$

should be a good estimator for $\theta = 4Nu$.

For example, there were $S_{11} = 14$ segregating sites in Kreitman's *ADH* sample. The denominator of the right-hand side of Equation 2.15 for $n = 11$ is 2.93, so the estimate of $4Nu$ is $\hat{\theta} = 4.78$. Theta for a nucleotide site, rather than for the entire locus, is $4.78/768 = 0.0062$. These estimates must be viewed with some skepticism because they require that the neutral model accurately reflect the evolutionary dynamics of the *ADH* locus and that the population be in equilibrium. Most population geneticists have reservations about the neutrality of replacement mutations, but many do accept the neutrality of silent mutations. Thus, you will frequently see θ estimated for silent variation only. In the case of *ADH*, this involves including only the 13 segregating sites with silent variation.

Problem 2.12 *Estimate θ for silent variation at the* ADH *locus.*

The coalescent may be used to derive the probability that two alleles different by origin are different by state, $\hat{\mathcal{H}}$. The two alleles will be different if a mutation occurred on the lineages leading from their common ancestor; otherwise they will be identical. As the lineages of the two alleles are traced backward in time, either a coalescence or a mutation will occur first. The probability that a coalescence occurs in any particular generation is $1/(2N)$, and the probability that a mutation occurs is

$$1 - (1-u)^2 \approx 2u.$$

The probability that a mutation is the first event to occur is just its relative probability of occurrence,

$$\frac{2u}{2u + 1/(2N)} = \frac{4Nu}{1+4Nu},$$

which is the same as Equation 2.9. This derivation of $\hat{\mathcal{H}}$ using a coalescent argument is both easier and more instructive than the difference equation approach used in Section 2.3 and hints at the power of the coalescent for solving ostensibly difficult problems. The coalescent has become an indispensable construct for the analysis sequence data in population genetics. Dick Hudson, one of the pioneers in the application of coalescent theory, has written an excellent paper describing its use (Hudson 1990).

2.7 Binomial sampling

The description of genetic drift in Section 2.2 was based on the probability of identity by state, $\mathcal{G}$. Although $\mathcal{G}$ is perfect for describing the average rate of decay of variability, it does not give a good feeling for the underlying randomness of genetic drift. In fact, the equation

$$\mathcal{H}_t = \mathcal{H}_0 \left(1 - \frac{1}{2N}\right)^t$$

could give the impression that the heterozygosity of any particular population decreases nonrandomly. Nothing could be further from the truth, as illustrated in Figure 2.8, which uses the same data as Figure 2.1 but plots heterozygosities rather than allele frequencies. Note that the heterozygosity of any particular population does not decrease monotonically. Rather, it jumps up and down, eventually hitting zero and staying there. After staring at Figure 2.8 for a while, it is reasonable to start wondering exactly what $\mathcal{H}_t$ really is. What does it correspond to in Figure 2.8? A promising conjecture is that $\mathcal{H}_t$ is the average heterozygosity of a very large number of replicate populations. Not a bad conjecture, but it is off by a factor of $1 - 1/(2N)$, as we shall see in this section.

The short algorithm used for the computer simulations at the beginning of this chapter also happens to be an accurate description of the way most

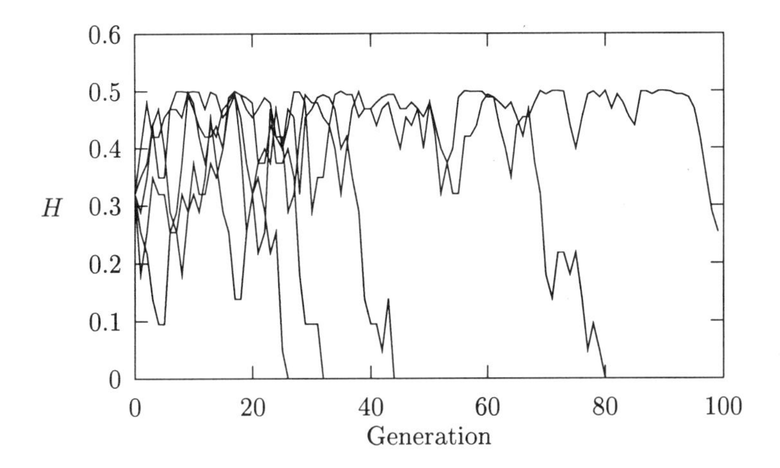

Figure 2.8: The heterozygosity for each of the five populations simulated by computer and illustrated in Figure 2.1.

population geneticists model genetic drift. The heart of the algorithm is the copying of alleles from the current generation into the next generation, which is repeated $2N$ times. The probability that i A_1 alleles make it into the next generation is the binomial probability

$$\text{Prob}\{i\ A_1 \text{ alleles}\} = \frac{(2N)!}{i!\,(2N-i)!} p^i q^{2N-i},$$

where i can be $0, 1, \ldots 2N$. The binomial distribution (see Appendix B) describes the probability of i successes in n independent experiments, where the probability of success in any one experiment is p. With genetic drift, the experiment is repeated $2N$ times and the probability of success—the probability of copying an A_1 allele—is the allele frequency p.

The mean of a binomial random variable with parameters n and p is np. Thus, the mean number of A_1 alleles to appear in the next generation is

$$E\{i\} = 2Np.$$

The allele frequency in the next generation is $i/2N$. The mean allele frequency is

$$E\{i/2N\} = E\{i\}/2N = p,$$

by Equation B.9. In other words, the mean allele frequency does not change under genetic drift.

A more complete description of the mean change in p is

$$E\{\Delta_N p \mid p\} = E\{i/2N - p \mid p\} = E\{i/2N \mid p\} - p = 0.$$

This stream of equalities uses a a new symbol, $|$, which stands for *given*, making the expectation a conditional expectation. For example, $E\{\Delta p \mid p\}$ is the mean change in p given its current value, p. We need this device because we cannot know with certainty what p will be in any particular generation as it is changing at random each generation. But, if we were told its value in a particular generation, then we could say something about its value in the next generation. The | symbol is shorthand for being told the current value of p.

The variance of the change in the allele frequency may be found by a similar argument,

$$\mathrm{Var}\{\Delta p \mid p\} = \mathrm{Var}\{i/2N - p \mid p\} = \frac{pq}{2N}.$$

The last step is interesting. First, note that the conditioning on p makes the expression $i/2N - p$ the difference between a random quantity, $i/2N$, and a nonrandom quantity, p. (p is not random because of the conditioning on its value.) From Equation B.10 we know that

$$\mathrm{Var}\{i/2N - p\} = \mathrm{Var}\{i\}/(2N)^2$$

when p is a constant. As $\mathrm{Var}\{i\} = 2Npq$, we obtain our important result

$$\mathrm{Var}\{\Delta p \mid p\} = \frac{pq}{2N}. \tag{2.16}$$

All of the above shows that the number of A_1 alleles in the daughter generation is binomially distributed and that the mean and variance of the change in the allele frequency are

$$E\{\Delta p\} = 0 \tag{2.17}$$

$$\mathrm{Var}\{\Delta p\} = \frac{pq}{2N}. \tag{2.18}$$

The fact that the variance is inversely proportional to the population size means that the dispersion of the allele frequency around the current value will be less in larger populations. The actual distributions of allele frequencies in the daughter generation for population sizes of 10 and 100 are illustrated in Figure 2.9. The narrow range of likely p values for the larger population size is apparent.

How can we connect this development of genetic drift with the previous one based on $\mathcal{G}$? Let's begin with some notation. Call the allele frequencies in the tth generation p_t and q_t and the homozygosity

$$G_t = p_t^2 + q_t^2.$$

Our conjecture that $\mathcal{G}_t$ is equal to the mean of G_t suggests that a natural place to begin our investigation is with the expected value of G_{t+1} given its value in generation t. Rather than conditioning on the current value of the homozygosity, it is somewhat easier to condition on the allele frequency in the

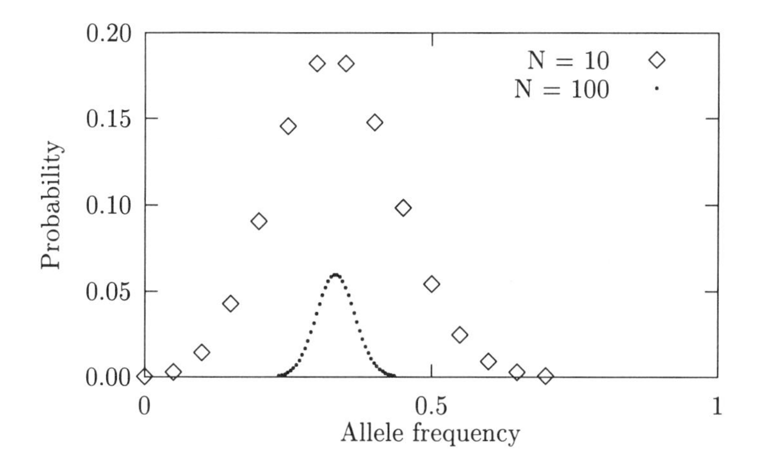

Figure 2.9: The probabilities of allele frequencies after one round of random mating for population sizes of $N = 10$ and $N = 100$ and initial allele frequency $p = 1/3$.

current generation, which amounts to the same thing,

$$\begin{aligned} E\{G_{t+1} \mid p_t\} &= E\{p_{t+1}^2 \mid p_t\} + E\{q_{t+1}^2 \mid q_t\} \\ &= \frac{p_t q_t}{2N} + p_t^2 + \frac{p_t q_t}{2N} + q_t^2 \\ &= \frac{1}{2N} + (1 - \frac{1}{2N})(p_t^2 + q_t^2) \\ &= \frac{1}{2N} + (1 - \frac{1}{2N})G_t. \end{aligned}$$

In verifying these steps, recall that

$$\text{Var}\{p\} = E\{p^2\} - E\{p\}^2$$

and that

$$2pq = 1 - p^2 - q^2.$$

At this junction, we have

$$E\{G_{t+1} \mid p_t\} = \frac{1}{2N} + (1 - \frac{1}{2N})G_t, \tag{2.19}$$

which is almost the same as Equation 2.1.

Equation 2.19 differs from Equation 2.1 in two ways. The first is that the quantity on the left side of Equation 2.19, $E\{G_{t+1} \mid p_t\}$, looks like a different beast than the analogous quantity on the right side, G_t. By contrast, $\mathcal{G}$ appears on both sides of Equation 2.1. The other difference is that both $E\{G_{t+1} \mid p_t\}$ and G_t are actually random variables that depend on the allele frequency in generation t, while $\mathcal{G}_{t+1}$ and $\mathcal{G}_t$ are both both nonrandom quantities. To make Equation 2.19 like Equation 2.1, we must, at the very least, remove the randomness, which can be accomplished by taking the expectation of both sides with respect to the distribution of the allele frequency at time t. Write the expected value of G_t with respect p_t as

$$E_{p_t}\{G_t\} = \overline{G}_t$$

and that of G_{t+1} as

$$\begin{aligned} E_{p_t}\{E\{G_{t+1} \mid p_t\}\} &= E\{E_{p_t}\{G_{t+1} \mid p_t\}\} \\ &= E\{\overline{G}_{t+1}\} \\ &= \overline{G}_{t+1}. \end{aligned}$$

The first line uses the fact that the order in which expectations are taken does not matter.* The second line uses the definition of $\overline{G}$. The third line is trivial as the expected value of a constant, $\overline{G}_{t+1}$ in this case, is just the constant. Substituting these results into Equation 2.19 gives

$$\overline{G}_{t+1} = \frac{1}{2N} + (1 - \frac{1}{2N})\overline{G}_t,$$

which is the same as Equation 2.1 and whose solution is

$$\overline{H}_t = 1 - \overline{G}_t = H_0 \left(1 - \frac{1}{2N}\right)^t. \tag{2.20}$$

We do not need a line over H_0 because the initial condition is a fixed quantity, so there is no need to take its mean.

Just because $\mathcal{G}_t$ and $\overline{G}_t$ satisfy the same equation does not mean that they are equal. The solution to a dynamical equation like Equation 2.1 depends on the initial value of the state variable. You should have solved Problem 2.6 by now and should know that

$$G_0 = \frac{1}{2N} + \left(1 - \frac{1}{2N}\right)\mathcal{G}_0,$$

or, equivalently, that

$$H_0 = \left(1 - \frac{1}{2N}\right)\mathcal{H}_0.$$

Plugging this into Equation 2.20 gives

$$\overline{H}_t = \mathcal{H}_0 \left(1 - \frac{1}{2N}\right)^{t+1},$$

which differs from the value of $\mathcal{H}_t$ given in Equation 2.3 by the factor $1 - 1/2N$. For large populations, this factor will be so close to one that $\mathcal{G}_t \approx \overline{G}_t$. Finally,

*Although in general expectations cannot be reversed, in our context they may be.

we are able to conclude with confidence that the expected heterozygosity of a population decreases at the rate $1/2N$.

While the amount of work needed to reach a rather simple conclusion might seem excessive, it is work well spent. Not only did we learn more about the stochastic nature of genetic drift, but we also put ourselves in a position to solve some more important problems, like the fixation probability of a selected mutation.

2.8 Answers to problems

2.1 The probability that a particular allele is not chosen on a single draw is $1 - 1/(2N)$. As each draw is with replacement, the probability that the allele is not drawn at all is $[1 - 1/(2N)]^{2N}$. For large populations, this probability approaches $e^{-1} \approx 0.37$, using the hint in the statement of the problem.

2.2 Here is a simulation written in the C programming language. In the declarations `int` means an integer data type and `double` means a double-precision floating-point data type. The `for` statement is a standard loop whose arguments give the initial value of the counter, the stop condition, and the change in the counter for each pass through the loop, respectively. The `++` operator increments the variable on its left by one. `(double)` before a variable converts the variable from an integer to a double. `random()` is a function returning an integer random number, and `RAND_MAX` is the maximum random integer. Thus, `random() / (double) RAND_MAX` evaluates to random numbers uniformly distributed between zero and one. (Different implementations of C may use different conventions for random numbers.)

```
#include <stdlib.h>

main() {
  /* declarations */
  int twoN = 40;
  double p = 0.2;
  int generation, number_of_A_1, i;
  /* the simulation */
  for (generation = 0; generation < 100; generation++) {
    number_of_A_1 = 0;
    for (i = 0; i < twoN; i++)
      if (random() / (double) RAND_MAX < p)
        number_of_A_1 = number_of_A_1 + 1;
    p = (double) number_of_A_1 / twoN;
    printf("%f\n", p);
  }
}
```

2.3 One approach is to simulate the process by flipping a coin and designating heads as the event that the individual in the next generation is a heterozygote and tails as the probability that it is a homozygote. The average number of flips until a tail appears is the same as the average number of generations until the population becomes homozygous. Alternatively, you can notice that the time to homozygosity is a geometric random variable and use the properties of geometric random variables as given in Appendix B to obtain the answer.

2.6 If sampling is with replacement, then the probability of choosing allele A_i on two successive draws from a given population is p_i^2. Thus, the probability of choosing two alleles that are identical by state is $\sum p_i^2$, which is the homozygosity of the population, G, as defined in Equation 1.2. G can be described in an entirely different way. Two alleles sampled with replacement will be identical by state if they are identical by origin, which occurs with probability $1/2N$. The probability that two alleles that are different by origin are identical by state is $\mathcal{G}$. Thus,

$$G = \frac{1}{2N} + \left(1 - \frac{1}{2N}\right)\mathcal{G}.$$

2.7 The number of mutations that are one step away is obtained by noting that there are 3000 sites where a mutation can occur and three nucleotides that can replace the original, giving $3000 \times 3 = 9000$ different mutations. For the number that are two steps away, note that there are 3000 sites for the first mutation and 2999 for the second mutation, so

$$3000 \times 2999 \times 3 \times 3 = 80{,}973{,}000$$

mutations are two steps away.

Chapter 3

Natural Selection

The results of natural selection, the evolutionary force most responsible for adaptation to the environment, are evident everywhere, yet it is remarkably difficult to observe the time course of changes brought about by selection. The reason, of course, is that most evolutionary change is extraordinarily slow. Significant changes in the frequencies of genotypes take longer than the lifetime of a human observer. This temporal imbalance is the greatest obstacle to the study of evolution and is the main reason why so much of our understanding of evolutionary processes comes from theoretical and mathematical arguments rather than direct observation, as is typical in other areas of biology.

Occasionally, we are able to observe natural selection in action either because the strength of selection is so great that change occurs very quickly or because the organism, perhaps a bacteria or virus, has a very short generation time. The European scarlet tiger moth, *Panaxia dominula*, provides one well-studied example. In a population just outside of Oxford, England, an allele that reduces the spotting on the forewing, the *medionigra* allele, is found in fairly high frequency. As this allele is found nowhere else, it has attracted attention from butterfly enthusiasts. The frequency of the *medionigra* allele declined fairly steadily from 1939 until 1955, after which it began hopping around erratically, as illustrated in Figure 3.1. Although the complete record is difficult to interpret, the period of steady decline appears to be a case of natural selection preferring the common allele over the *medionigra* allele. If so, how strong is the selection? Other questions come to mind as well. Why is the *medionigra* allele less fit? If it is less fit, how did it get to a frequency of 10 percent before beginning its decline? While we will not be able to provide complete answers to any of these questions, we will be able to discuss them much more intelligently after a theoretical investigation of the nature and consequences of natural selection.

In this chapter, we will discover how natural selection changes allele frequencies by examining some one-locus models of selection. Natural selection works when genotypes have different fitnesses. To a geneticist, fitness is just another trait with a genetic component. To an evolutionist, it is the ultimate trait because it is the one upon which natural selection acts. Fitness is a complicated

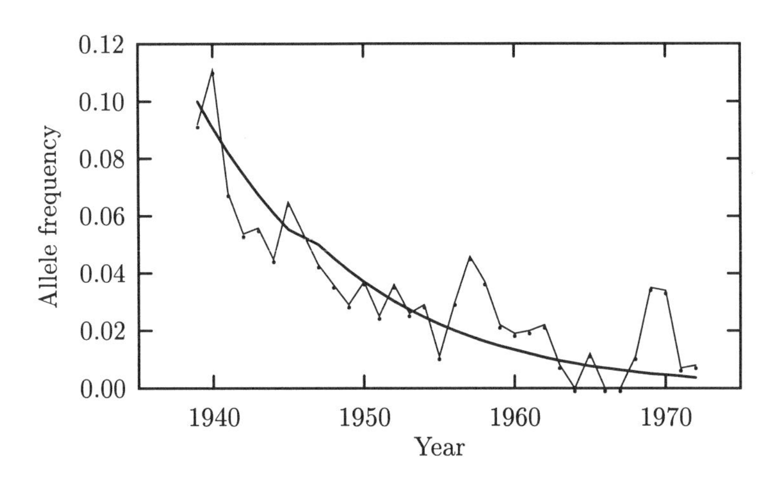

Figure 3.1: The observed frequency of the *medionigra* allele in the scarlet tiger moth population compared to the expected frequency assuming a 10 percent disadvantage.

trait, even in the context of a simple one-locus, two-allele model. There is individual fitness, genotype fitness, relative fitness, and absolute fitness. We will spend some time making these different aspects of fitness clear before tackling the problem of the dynamics of natural selection.

An examination of the dynamics of natural selection quickly leads to the conclusion that the dominance relationships between alleles affecting fitness have a profound affect on the outcome of selection. Fortunately, the dominance of fitness alleles can be investigated experimentally; in Section 3.5 a study of viability in *Drosophila melanogaster* populations is described. A major conclusion of this study is that there is an inverse homozygous-heterozygous effect for deleterious alleles: alleles that have large deleterious effects when homozygous tend to be nearly recessive, whereas alleles with small homozygous effects tend to be nearly additive. A casualty of the study is overdominance, the form of dominance that is often invoked to explain selected polymorphisms. However, the subsequent section shows that selection in a variable environment can promote polymorphism even when heterozygotes are intermediate in fitness.

Section 3.7 examines the interaction of genetic drift and selection. Genetic drift has a major influence on the fate of rare alleles even in very large populations. In fact, the sad fate of most advantageous mutations is extinction, which leads to the view that evolution is fundamentally a random process that is not repeatable or reversible. Finally, a version of the neutral theory that involves the fixation of deleterious alleles is described.

This is an ambitious chapter, with many more new topics than were in the preceding chapters. It is also more difficult than the previous material because the mathematics of selection lacks the elegance and simplicity of the mathematics of drift and mutation.

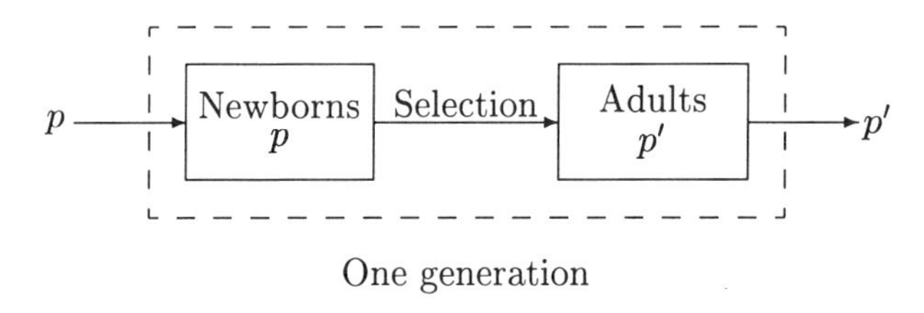

Figure 3.2: The simple life cycle used in the fundamental model of selection.

3.1 The fundamental model

Natural selection is most easily studied in the context of an autosomal locus in a hermaphroditic species whose life cycle moves through a synchronous cycle of random mating, selection, random mating, selection, and so forth. Our entrance into the cycle is with newborns produced just after a round of random mating by their parents. Figure 3.2 shows that the frequency of the A_1 allele among the newborns is called p, which is the same as the allele frequency in their parents. As their parents mated at random, the genotype frequencies of the newborns will conform to Hardy-Weinberg expectations.

The newborns must survive to adulthood in order to reproduce. The probability of survival of an individual will, in general, depend on the genotype of the individual. Let the probabilities of survival or, as they are more usually called, the viabilities of A_1A_1, A_1A_2, and A_2A_2 individuals be w_{11}, w_{12}, and w_{22}, respectively. Viabilities may be thought of as either probabilities of survival of individuals or the fraction of individuals that survive. The latter allows us to see immediately the consequences of selection because the frequency of a genotype after selection is proportional to its frequency before selection times its viability, or

$$\text{frequency after selection} \propto \text{newborn-frequency} \times \text{viability}.$$

For example, the frequency of A_1A_1 in the adults is proportional to p^2w_{11}.

To obtain the relative frequencies of the three genotypes in the adults, we must find a constant of proportionality such that the sum of the three genotype frequencies in the adults is one. The following worksheet shows how this is done.

Genotype:	A_1A_1	A_1A_2	A_2A_2
Frequency in newborns:	p^2	$2pq$	q^2
Viability:	w_{11}	w_{12}	w_{22}
Frequency after selection:	$p^2w_{11}/\bar{w}$	$2pqw_{12}/\bar{w}$	$q^2w_{22}/\bar{w}$.

The constant of proportionality,

$$\bar{w} = p^2w_{11} + 2pqw_{12} + q^2w_{22},$$

is chosen such that

$$\frac{p^2w_{11}}{\bar{w}} + \frac{2pqw_{12}}{\bar{w}} + \frac{q^2w_{22}}{\bar{w}} = 1,$$

as required. The quantity $\bar{w}$ has special meaning in population genetics. It is called the mean fitness of the population. (If the concept of a mean is unfamiliar, read from the beginning of Appendix B through page 157.)

After selection, the frequency of the A_1 allele may have changed. The new allele frequency, p', is

$$p' = \frac{p^2 w_{11} + pq w_{12}}{\bar{w}}.$$

(Don't forget that each heterozygote has only one A_1 allele.) The change in the frequency of the A_1 allele in a single generation, $\Delta_s p = p' - p$, follows from

$$\begin{aligned} p' - p &= \frac{p^2 w_{11} + pq w_{12} - p\bar{w}}{\bar{w}} \\ &= \frac{p[q w_{11} + q(1-2p) w_{12} - q^2 w_{22}]}{\bar{w}}, \end{aligned}$$

which simplifies to

$$\boxed{\Delta_s p = \frac{pq[p(w_{11} - w_{12}) + q(w_{12} - w_{22})]}{p^2 w_{11} + 2pq w_{12} + q^2 w_{22}}} \tag{3.1}$$

This is probably the single most important equation in all of population genetics and evolution! Admittedly, it isn't pretty, being a ratio of two polynomials with three parameters each. Yet, with a little poking around, this equation easily reveals a great deal of the dynamics of natural selection.

Problem 3.1 *In 1940, the frequency of the* medionigra *allele in the Oxford population was about* $p = 0.1$. *If the viabilities of the three genotypes were* $w_{11} = 0.9$, $w_{12} = 0.95$, *and* $w_{22} = 1$, *what would be the frequency of* medionigra *in the newborns of 1941?*

3.2 Relative fitness

Notice that the terms in the numerator and denominator of Equation 3.1 all have a viability as a factor. Thus, if we were to divide the numerator and denominator by a viability, say by w_{11}, every viability in Equation 3.1 would become a ratio of that viability and w_{11}, yet the numerical value of Δp would not change at all. In other words, we could use as our definition of viability either the original definition based on absolute viabilities or a new one based on the relative viabilities of genotypes when compared to one particular genotype.

Genotype:	A_1A_1	A_1A_2	A_2A_2
Viability	w_{11}	w_{12}	w_{22}
Relative viability:	1	w_{12}/w_{11}	w_{22}/w_{11}

In either case, the dynamics of selection as captured in Equation 3.1 are the same. An important insight in its own right, this is also of great utility as it allows a much more informative choice of parameters than the w_{ij}'s used thus far.

Up to this point, w_{ij} has been called the viability of genotype A_iA_j. More often, w_{ij} is called the fitness, or sometimes the absolute or Darwinian fitness, of genotype A_iA_j. In nature, the fitness of a genotype has many components including viability, fertility, developmental time, mating success, and so forth. Most of these components, other than viability, cannot be included in a simple model like that defined by Equation 3.1. Yet, if the differences in fitnesses between genotypes are small, Equation 3.1 is a good approximation to the actual dynamics as long as the values of the w_{ij} are chosen appropriately. Here, we will not investigate these more complicated models but will from this point on refer to w_{ij} as a fitness and allow it to take on any values greater than or equal to zero. As the dynamics of selection depend on relative fitnesses, nothing really changes by allowing this broadened scope for w_{ij}.

A common notational convention for relative fitnesses is

Genotype:	A_1A_1	A_1A_2	A_2A_2
Relative fitness:	1	$1-hs$	$1-s$

where $1-hs = w_{12}/w_{11}$ and $1-s = w_{22}/w_{11}$.

The parameter s is called the selection coefficient. It is a measure of the fitness of A_2A_2 relative to that of A_1A_1. If the selection coefficient is positive, A_2A_2 is less fit than A_1A_1; if it is negative, A_2A_2 is more fit. In most of what follows, we will assume that the selection coefficient is in the range $0 \leq s \leq 1$. Nothing is lost in doing this because the labeling of alleles is completely arbitrary. The symbol A_1 will usually be attached to the allele whose homozygote is more fit than the other homozygote.

The parameter h is called the heterozygous effect. It is a measure of the fitness of the heterozygote relative to the selective difference between the two homozygotes. As such, it is really a measure of dominance, as shown in the following table.

$h = 0$	A_1 dominant, A_2 recessive
$h = 1$	A_2 dominant, A_1 recessive
$0 < h < 1$	incomplete dominance
$h < 0$	overdominance
$h > 1$	underdominance

Only the cases of incomplete dominance, overdominance, and underdominance are of general evolutionary interest. The cases of complete dominance given in the first two lines of the table are regarded as special cases that are unlikely to occur for any pair of naturally occurring alleles.* Even classic cases of complete dominance like "recessive" lethals in human populations are now thought to be cases of incomplete dominance with $s = 1$ and h small, say about 0.01, but definitely greater than zero. The case $h = 1/2$ is of evolutionary importance

*Cases of complete dominance abound for morphologic traits. Our interest, however, is only in the effects of alleles on fitness.

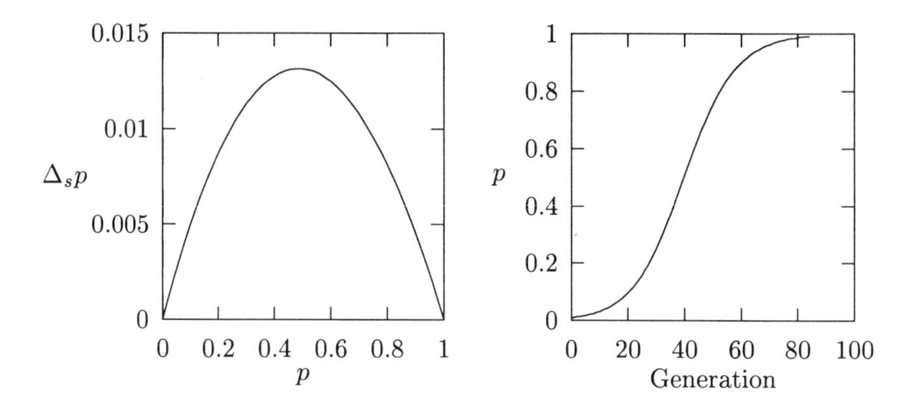

Figure 3.3: Properties of directional selection with $h = 0.5$ and $s = 0.1$. The left-hand graph shows the change in the allele frequency in a single generation. The right-hand graph shows the evolution of the allele frequency over 100 generations.

as many alleles with a very small effect on fitness are close to additive; that is, close to the situation where the heterozygote is exactly intermediate between the two homozygotes.

Problem 3.2 *If the fitnesses of the genotypes A_1A_1, A_1A_2, and A_2A_2 are 1.5, 1.1, and 1.0, respectively, what are the values of the selection coefficient and the heterozygous effect?*

Problem 3.3 *Fitnesses must always be greater than or equal to zero. What limitations does this place on the values of h for a given value of s, assuming that $s \geq 0$?*

Equation 3.1 for the single-generation change in the frequency of the A_1 allele, using relative frequencies, becomes

$$\Delta_s p = \frac{pqs[ph + q(1-h)]}{\bar{w}}. \tag{3.2}$$

(To obtain this equation, simply make the substitutions $w_{11} = 1$, $w_{12} = 1 - hs$, and $w_{22} = 1 - s$ in equation 3.1.) The mean fitness becomes

$$\bar{w} = 1 - 2pqhs - q^2 s \tag{3.3}$$

when using relative fitnesses. Not only is Equation 3.2 simpler than Equation 3.1, it is also more suggestive of the dynamics. We will soon see that h determines where the allele frequency ends up and s determines how quickly it gets there.

Thus far, we have discussed absolute and relative fitness, but not individual fitness, which is, as its name implies, a property of an individual as opposed to a genotype. If the only factor determining fitness were viability, then an individual's fitness would be either zero or one, depending on whether or not the individual survived to reproduce. The absolute fitness of a particular genotype is simply the average of the individual fitnesses of all those individuals with that genotype. For example, if the probability that an A_2A_2 individual's fitness is one equals 0.9 and is zero equals 0.1, then the absolute fitness of the A_2A_2 genotype is

$$0.9 \times 1 + 0.1 \times 0 = 0.9.$$

3.3 Three kinds of selection

Equation 3.2 may be used to solve the following problem: Given the initial frequency of the A_1 allele and the parameters s and h, what will be the ultimate fate of the A_1 allele in the population? Will it take over in the population ($p \to 1$), disappear from the population ($p \to 0$), approach some intermediate value ($p \to \hat{p}$), or not change at all? As we shall see, all four outcomes are possible. Which one prevails depends on the dominance relationships between alleles and on the initial frequency of the allele.

The type of selection that Darwin had in mind in *On the Origin of Species* (1859) is called directional selection. Directional selection occurs with incomplete dominance ($0 < h < 1$). The fitness of A_1A_1 exceeds that of A_1A_2, which, in turn, exceeds that of A_2A_2. It should come as no surprise that p continually increases or, equivalently, that $\Delta_s p > 0$. The change in the allele frequency in a single generation, $\Delta_s p$, as a function of p is illustrated in the left-hand side of Figure 3.3. Because $s > 0$, Equation 3.2 shows that the sign of $\Delta_s p$ is determined by

$$ph + q(1 - h),$$

which is always positive with incomplete dominance ($0 < h < 1$). Thus, the allele frequency always increases no matter what its current value and, as a consequence, $p \to 1$, as illustrated in the right-hand frame of Figure 3.3. The rate of change of p is strongly dependent on p itself. Evolution by natural selection proceeds very slowly when there is little genetic variation; that is, when p is close to zero or one. Selection is most effective when genetic variation is near its maximum, $p = 1/2$.

Problem 3.4 *Graph $\Delta_s p$ for $s = 0.1$ with $h = 0.1$ and $h = 0.9$. How does dominance affect directional selection?*

Note that p will not equal one in a finite number of generations. As p gets closer to one, $\Delta_s p$ approaches zero fast enough to prevent $p = 1$ in finite time. However, in finite populations genetic drift dominates selection when p gets very close to one and will thus cause fixations to occur in finite time.

Problem 3.5 *Write down Equation 3.2 for the special case $h = 1/2$. Find the allele frequencies in two successive generations when the initial value of p is 0.1 and $s = 0.1$. If you can program a computer, continue this for 200 generations and graph the result.*

The decrease in the frequency of the *medionigra* allele in *Panaxia* from 1939 until about 1955 may well be due to directional selection. If the selection coefficient for the *medionigra* allele were $s = 0.1$, the heterozygous effect $h = 1/2$, and the initial frequency $q = 0.1$, then the allele frequency trajectory obtained from Equation 3.2 fits the observed allele frequency quite well, as illustrated in Figure 3.1. Does this constitute a "proof" that direction selection is responsible for the allele frequency change? Unfortunately, no. Many different forms of selection have trajectories that fit the data as well as does directional selection. A convincing demonstration that directional selection is operating requires direct measurements of the selection coefficient to show that $0 < s < 1$ and of the heterozygous effect to show that $0 < h < 1$. The obstacles to obtaining these estimates in a natural setting are almost insurmountable. Not only are there the obvious problems of determining the viability, fertility, mating success, longevity, and so on of genotypes with sufficient accuracy, there is also the problem of demonstrating that the measured parameters are due to alleles at the *medionigra* locus and not to those at some closely linked locus. A good introduction to the problems of measuring selection in nature is John Endler's book, *Natural Selection in the Wild* (1986).

The entire history of the *medionigra* allele is a mystery. The allele is found only in the Oxford population, where it appears to have been at a selective disadvantage from at least 1939 to 1955. If it were always at a disadvantage, then why was its frequency as high as 10 percent in 1939? An obvious answer, one that fits with the standard Darwinian view of the world, is that the environment has changed. The fitness of a genotype is determined by its adaptation to the environment in which it finds itself. Even a casual observer notices that environments are constantly changing. The physical environment changes on many different time scales from seconds to million of years. There are daily and seasonal temperature cycles; ice ages are temperature cycles with a period of tens of thousands of years. Continental drift causes major climatic changes on even longer time scales. More subtle are the changes in the biological environment. Most creatures are both predator and prey; all, save perhaps viruses, are attacked by pathogens. These components of the biological environment are just as variable as the physical environment, perhaps even more so. Our picture of evolution should never be one of constant improvement in a static environment, but rather a desperate evolutionary race to avoid extinction in a constantly deteriorating environment. With this view, it is perfectly natural to assume that *medionigra* was more fit in the Oxford environment at some time prior to 1939 and less fit after 1939. Of course, we have no way to reconstruct history to find out if this is so. Nor can we rule out such possibilities as a sudden increase in the mutation rate due a transposable element or some other unorthodox mutational event.

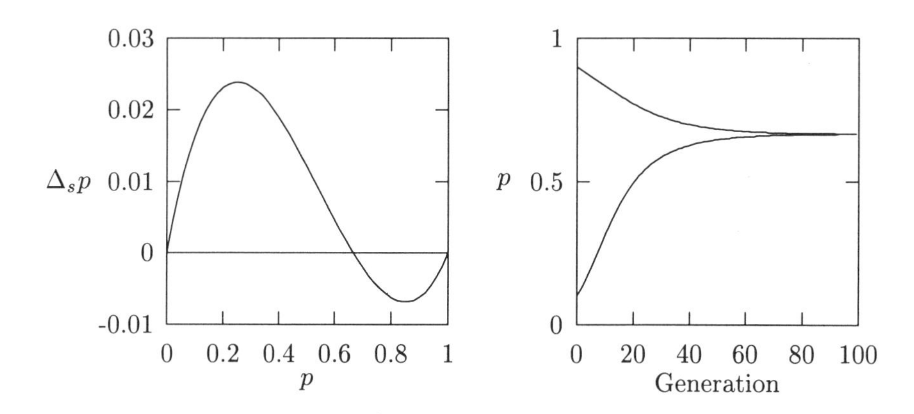

Figure 3.4: Properties of balancing selection with $h = -0.5$ and $s = 0.1$.

The second type of selection occurs when there is overdominance ($h < 0$). In this case, the allele frequency approaches an equilibrium value, $p \to \hat{p}$, as is readily inferred from the graph of $\Delta_s p$ versus p illustrated in the left-hand frame of Figure 3.4. When p is close to zero, $\Delta_s p > 0$ and the allele frequency will increase. When p is close to one, $\Delta_s p < 0$ and p will decrease. Therefore, p must approach an equilibrium that is between zero and one. The equilibrium is that point where the allele frequency no longer changes, $\Delta_s p = 0$. From Equation 3.2, we see that this occurs when

$$\hat{p}h + (1 - \hat{p})(1 - h) = 0,$$

or

$$\hat{p} = \frac{h - 1}{2h - 1}. \tag{3.4}$$

As both alleles are kept in the population in a balanced or stable equilibrium, this form of selection is called balancing selection.

As a technical aside, the method used to determine the existence of an internal equilibrium is called end-point analysis. This technique is very useful for describing the qualitative behavior of complicated dynamical models when the global behavior is too difficult to understand. End-point analysis could have been used for directional selection as well. There, the allele frequency increases when both rare and common, suggesting that $p = 1$ is a stable equilibrium and $p = 0$ is an unstable equilibrium.

Problem 3.6 *Graph $\hat{p}$ as a function of h for $-1 < h < 2$. Locate the regions of the graph that correspond to incomplete dominance and overdominance, and discuss the values of $\hat{p}$ in these two regions.*

The most thoroughly studied example of overdominance is the sickle-cell hemoglobin polymorphism found in many human populations in Africa. Hemoglobin, the oxygen-carrying red protein found in red blood cells, is a tetramer composed of two alpha chains and two beta chains. In native West and Central African populations, the S allele of beta hemoglobin reaches a frequency as high as 0.3 in some areas. The more common A allele is found at very high frequency in most other areas of the world. The two alleles differ only in that the S allele has a glutamic acid at its sixth amino position while the A allele has a valine. The glutamic acid causes the hemoglobin to form crystal aggregates under low partial pressures of oxygen, as occur, for example, in the capillaries. As a result, SS homozygotes suffer from sickle-cell anemia, a disease that is often fatal.

The S allele could not have reached a frequency of 0.3 unless AS heterozygotes are more fit than AA homozygotes. This is precisely the case in regions where malaria is endemic, for there the heterozygotes are somewhat resistant to severe forms of malaria. The resistance is due to the sickling phenomena, which makes red blood cells less suitable for *Plasmodium falciparum.* In an old study from 1961, it was shown that the viability of AS relative to AA is 1.176 in regions with malaria. Assuming that the fitness of SS is zero ($s = 1$), $h = -0.176$. Plugging this into Equation 3.4 gives $\hat{p} = 0.87$ or $\hat{q} = 0.13$ for the S allele, which is nestled right in the middle of allele frequencies in regions with endemic malaria.

The complete story of selection on beta hemoglobin variation is more complicated than the bit presented here. A very readable, though somewhat dated, account may be found in *The Genetics of Human Populations* by Lucca Cavalli-Sforza and Walter Bodmer (1971). Even our abbreviated account emphasizes once again that selection occurs in an environment that is always changing. The relevant environment for the sickle-cell polymorphism is the biological environment embodied in falciparum malaria. Populations of *Plasmodium* fluctuate in both time and space with consequent changes in the fitnesses of the beta hemoglobin genotypes. In areas without malaria, the form of selection shifts from balancing selection to directional selection.

The final form of selection, disruptive selection, occurs when there is underdominance ($h > 1$). A graph of $\Delta_s p$ versus p in this case shows that p will decrease when rare and increase when near one. (You will be asked to draw this graph in Problem 3.7) That's strange: The outcome of selection depends on the initial frequency of the allele! In fact, the allele frequency will approach zero if the initial value of p is less than $\hat{p}$, where $\hat{p}$ is given by Equation 3.4. The allele frequency will approach one if the initial value of p is greater than $\hat{p}$. If, by some bizarre chance, $p = \hat{p}$, the allele frequency will not change at all. $\hat{p}$ is an unstable equilibrium because the smallest change in p will cause the allele frequency to move away from $\hat{p}$. A small change in p might well be caused by genetic drift.

Problem 3.7 *Graph $\Delta_s p$ versus p for an underdominant locus. Use the figure to convince yourself that the description of disruptive selection given in the preceding paragraph is correct.*

There are very few, if any, examples of underdominant alleles in high frequency in natural populations. However, the fact that closely related species sometimes have chromosomes that differ by inversions or translocations suggests that underdominant chromosomal mutations do occasionally cross the unstable equilibrium. The evolutionary forces that push the frequencies over the unstable point are not known, although both genetic drift and meiotic drive are likely candidates.

There is something unsatisfying about the description of the three forms of natural selection. They come off as a series of disconnected cases. One might have hoped for some unifying principle that would make all three cases appear as instances of some more general dynamic. In fact, Sewall Wright found unity when he wrote Formula 3.2 in the more provocative form

$$\Delta_s p = \frac{pq}{2\bar{w}} \frac{d\bar{w}}{dp}. \tag{3.5}$$

(The symbol $d\bar{w}/dp$ is the derivative or slope of the mean fitness viewed as a function of the allele frequency p.) Equation 3.5 shows that $\Delta_s p$ is proportional to the slope of the mean fitness function. If the slope is positive, then so is $\Delta_s p$. As a result, selection will increase p and, because $d\bar{w}/dp > 0$, will increase the mean fitness of the population. If the slope is negative, p will decrease and, because $d\bar{w}/dp < 0$, the mean fitness will increase once again. In other words, the allele frequency always changes in such a way that the mean fitness of the population increases. Moreover, the rate of change in p is proportional to the genetic variation in the population as measured by pq. Although we will not show it, the rate of change of the mean fitness, $\bar{w}$, is proportional to pq as well. Thus, selection always increases the mean fitness of the population and does so at a rate that is proportional to the genetic variation.

R. A. Fisher made a similar observation at about the same time as did Wright and called it the Fundamental Theorem of Natural Selection (Fisher 1958). Fisher showed that the change in the mean fitness is proportional to the additive genetic variation in fitness. (We will learn about the additive variance in Chapter 5.) As variances are always positive, the mean fitness will always increase when natural selection changes the allele frequency.

The Fundamental Theorem of Natural Selection is undeniably true for theoretical populations with simple selection at a single locus. However, with more loci or if fitness depends on the frequencies of genotypes or if it changes through time, the Fundamental Theorem no longer holds. Thus, it is neither fundamental nor a theorem; some have claimed that it has little to do with natural selection. Its biological significance has always been controversial. Yet, the metaphor suggested by the theorem that natural selection always moves populations upward on the "adaptive landscape" has proven to be a convenient one for simple descriptions of evolution without mathematics or deep understanding. The metaphor is stretched too far when applied to evolution for more than a few generations, as the change in fitness due to environmental change renders the metaphor inappropriate. Imagine climbing a mountain that keeps moving;

despite your best efforts, the peak remains about the same distance ahead. That is the proper metaphor for evolution.

Problem 3.8 *Graph the mean fitness of the population as a function of* p *for* $s = 0.1$ *and* $h = -0.5$, 0.5, *and* 1.5. *Do the peaks correspond to the outcomes of selection described above?*

3.4 Mutation-selection balance

The vast majority of mutations of large effect are deleterious and incompletely dominant. They enter the population by mutation and are removed by directional selection. A balance is reached where the rate of introduction of mutations is exactly matched by their rate of loss due to selection. The equilibrium number of deleterious mutations is large enough to have a major effect on many evolutionary processes. Among these are the evolution of sex and recombination and the avoidance of inbreeding. Most of these mutations are partially recessive, $h < 1/2$, so their effects are not always apparent unless the population is made homozygous either by genetic drift or by inbreeding. In this section, we will study the balance between mutation and selection and then go on in the next section to describe the dominance relationships between naturally occurring alleles.

Following our labeling conventions, A_2 will represent the deleterious allele whose frequency is increased by mutation and decreased by directional selection. Selection will be assumed to be sufficiently strong so that the frequency of A_2 is very small. As a consequence, the most important effect of mutation is to convert A_1 alleles into A_2 alleles. The reverse happens as well, but has little influence on the dynamics and can be ignored. Suppose, therefore, that there is one-way mutation from allele A_1 to allele A_2,

$$A_1 \xrightarrow{u} A_2,$$

where u is the mutation rate, the probability that a mutation from A_1 to A_2 appears in a gamete.

The effects of mutation on p may be described in the same way as was done in the discussion of the balance between mutation and genetic drift. For an allele in the next generation to be A_1, it must have been A_1 in the current generation and it must not have mutated,

$$p' = p(1 - u).$$

The change in p in a single generation is

$$\Delta_u p = -up. \tag{3.6}$$

Mutation rates are usually very small: 10^{-5} for visible mutations at a typical locus in *Drosophila* to 10^{-9} for a typical nucleotide. Thus, the frequency of A_1 decreases very slowly while the frequency of A_2 increases very slowly. If selection against A_2 is sufficiently strong, it will keep the frequency of A_2 very low, allowing the approximation

$$\Delta_u p = -u + qu \approx -u, \tag{3.7}$$

because $q \approx 0$.

Problem 3.9 *Follow the exact and approximate frequencies of the A_2 allele for two generations when the initial frequency of A_2 is zero and $u = 10^{-5}$. What is the relative error introduced by the approximation 3.7? (The relative error is the difference between the exact and approximate values divided by the exact value.)*

From Equation 3.2, we can write the change in the frequency of A_1 due to selection acting in isolation, when $q \approx 0$, as

$$\Delta_s p = \frac{pqs[ph + q(1-h)]}{1 - 2pqhs - q^2 s} \approx qhs. \tag{3.8}$$

The approximation is valid when $q \approx 0$, which implies that $p \approx 1$ and $\bar{w} \approx 1$.

At equilibrium, the change in the frequency of A_1 by mutation must balance the change due to selection,

$$\begin{aligned} 0 &= \Delta_u p + \Delta_s p \\ &\approx -u + qhs, \end{aligned}$$

which gives the equilibrium frequency of A_2,

$$\boxed{\hat{q} \approx \frac{u}{hs}} \tag{3.9}$$

The equilibrium frequency of a deleterious allele is approximately equal to the mutation rate to the allele divided by the selection against the allele in heterozygotes. Recall from Chapter 1 that rare alleles are found mainly in heterozygotes, not homozygotes. Thus, it is not surprising that the equilibrium frequency of deleterious alleles depends on their fitness in heterozygotes rather than homozygotes.

Deleterious alleles cause problems for populations. One measure of these problems is the genetic load of the population,

$$L = \frac{w_{\max} - \bar{w}}{w_{\max}}, \tag{3.10}$$

where $w_{\max}$ is the fitness of the maximally fit genotype in the population. The closer the mean fitness of the population is to the fitness of the most fit genotype, the less is the genetic load.

The mean fitness of a population at equilibrium under the mutation-selection balance is

$$\begin{aligned} \bar{w} &= 1 - 2\hat{p}\hat{q}hs - \hat{q}^2 s \\ &\approx 1 - 2\hat{q}hs \\ &\approx 1 - 2u. \end{aligned}$$

Remarkably, the presence of deleterious mutations decreases the mean fitness by an amount, $2u$, that is independent of the strength of selection in heterozygotes. The genetic load in this case is simply

$$L = \frac{1 - (1 - 2u)}{1} = 2u, \tag{3.11}$$

which follows from Equation 3.10 with $w_{\max} = 1$. When selection is weak, the frequency of the deleterious allele will be higher, but the detrimental effect of each allele on the mean fitness of the population is slight. When selection is strong, the frequency of A_2 is less, but the effect is greater. Hence, the independence of the load on the strength of selection follows. The biological significance of genetic load, like that of the Fundamental Theorem of Natural Selection, has been hotly debated over the years. Load is most useful when discussing deleterious alleles of measurable effect but is of dubious value for variation maintained by balancing selection or for directional selection of advantageous alleles.

Problem 3.10 *Derive the genetic load for an overdominant locus at equilibrium. (Do not include mutation.) Is this greater or less than the load of a population made up entirely of A_1A_1 individuals (and for which the A_2 allele does not exist, even as a possibility)? What are the implications of your answers on the biological significance of genetic loads?*

3.5 The heterozygous effects of alleles

In 1960, Rayla Greenberg and James F. Crow published a landmark paper reporting the results of a study "undertaken in an attempt to determine whether the effects of recurrent mutation on the population and the deleterious effects of inbreeding are due primarily to a small number of genes of major effect or to the cumulative activity of a number of genes with individually small effects." The study did this and considerably more. Of particular interest was its suggestion that mutations of large effect are almost recessive ($s \approx 1$ and $h \approx 0$) while those of small effect are almost additive ($s \approx 0$ and $h \approx 1/2$). The prevailing view in the late 1950s was that most deleterious mutations are completely recessive ($h = 0$). This paper is not only historically and scientifically important, but also pedagogically valuable because it uses many of the ideas developed in this and the previous chapters. In addition, it introduces an experimental methodology that is central to population genetics.

The design of the experimental part of the Greenberg and Crow paper was developed in the late 1930s by Alfred Sturtevant and Theodosius Dobzhansky. Back then, most of the genetic variation affecting fitness was thought to be due to rare, recessive, deleterious alleles. As rare alleles are usually heterozygous, these mutations would not be expressed in wild-caught individuals. An obvious way to study this "hidden variation" is to make individuals homozygous, as this allows the expression of recessive mutations. *Drosophila* was the only suitable organism for such a study because it alone allowed the experimental manipulation of entire chromosomes on which recombination is suppressed.

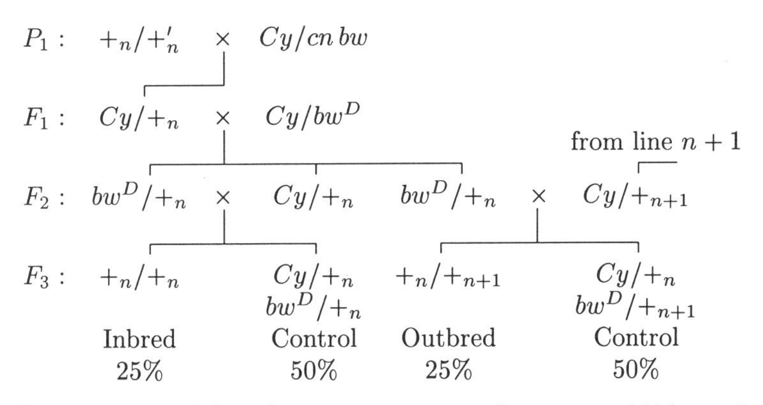

Figure 3.5: The *Drosophila melanogaster* crosses used to uncover hidden variation. In each cross, the male is on the left.

The experimental design is illustrated in Figure 3.5. The purpose of the design is to construct flies that are homozygous for their entire second chromosomes. The viabilities of these flies are then compared to those whose two second chromosomes are drawn independently from nature, thus mimicking random mating. The first class of flies will be called inbred, and the second class will be called outbred.* The details of the design are as follows:

P_1 In the parental generation, $+_n/+'_n$ represents one male fly, obtained from nature or from an experimental population, that initiates the nth line of crosses. The symbols $+_n$ and $+'_n$ stand for the two second chromosomes found in the male fly. One of the second chromosomes of this fly will ultimately be made homozygous. As there is no recombination in male *Drosophila*, the chromosomes in this original male remain intact. The male is crossed to a $Cy/cn\,bw$ female. Cy is a dominant second chromosome mutation, Curly wing, that is placed on a chromosome with one paracentric inversion on each arm to block recombination. The other chromosome has two recessive mutations, cinnabar eyes (cn) and brown eyes (bw). This initial cross is repeated 465 times, each repetition using an independently obtained male. Each repetition is called a line; the lines are numbered sequentially from 1 to 465.

F_1 A single $Cy/+_n$ male from each line is crossed to a Cy/bw^D female. The female Cy chromosome in this step is a slightly fancier version of the previous Cy chromosome, with a pericentric inversion, $SM1$, providing extra safeguards against recombination. The homolog to Cy in the female contains a dominant brown-eye mutation, bw^D. This is the critical step in

*Greenberg and Crow called members of these two classes *homozygotes* and *heterozygotes*, respectively. However, while *homozygotes* is accurate for the first class, the loci of the second class can be either homozygous or heterozygous at each locus as they are in Hardy-Weinberg proportions.

the design as it assures that only a single wild-caught second chromosome is used.

F_2 Two different crosses occur in this generation. The first is a mating of a a $bw^D/+_n$ male to a $Cy/+_n$ female from the same line (a brother-sister mating). The second is a cross of a $bw^D/+_n$ male to a $Cy/+_{n+1}$ female from the $(n+1)$st line.

F_3 The offspring from the brother-sister F_2 cross will fall into four classes: $+_n/+_n$, $Cy/+_n$, $bw^D/+_n$, and Cy/bw^D, which are easily recognized because Cy and bw^D are both dominant mutations. According to Mendel's law of segregation, these four classes should be equally frequent. However, as the Cy/bw^D flies are not used in the analysis, they are not included in Figure 3.5. The $+_n/+_n$ flies are homozygous at every locus on their second chromosome and for this reason are called inbreds. The offspring from the interline cross have the same phenotypic classes as those from the intraline cross, but the wild-type flies will contain two independently derived second chromosomes. They are formally the same as flies produced by random mating from the original population, so they are called outbreds.

The $Cy/+_n$ and $bw^D/+_n$ flies, called controls in the figure, are not inbred and are relatively vigorous. They are used as a reference for determining the relative viability of the $+_n/+_n$ flies according to the formula

$$+_n/+_n \text{ viability} = \frac{2 \times \text{number of } +_n/+_n \text{ flies}}{\text{numbers of } Cy/+_n \text{ and } bw^D/+_n \text{ flies}}.$$

The determination of the viability of outbred flies is done in a similar fashion.

Figure 3.6 illustrates the number of lines in each relative viabilities class among the inbred and outbred flies obtained after carrying out the crosses in Figure 3.5. Viability class 0 is made up of all lines with relative viability in the interval $0 \le v < 0.1$, class 0.1 includes those in the range $0.1 \le v < 0.2$, and so forth. The most striking aspect of the figure is the bimodal distribution of viabilities in the inbred flies. The left mode is due to "recessive" lethal mutations found on 106 of the 465 chromosomes examined. In other words, about 23 percent of all second chromosomes carry at least one mutation that is lethal in the homozygous state. As a single second chromosome represents about 20 percent of the total genes in *Drosophila*, it follows that most individuals carry at least one lethal mutation. *Drosophila* is not unusual in this regard. Most diploid organisms, including ourselves, carry a similar number of lethal mutations.

Problem 3.11 *Assuming that a single second chromosome contains 20 percent of the loci, calculate the probability that an individual will be completely free of lethal mutations.*

The graphed viabilities of inbred flies dips into a deep valley separating a left-hand lethal peak from the much broader right-hand deleterious peak.

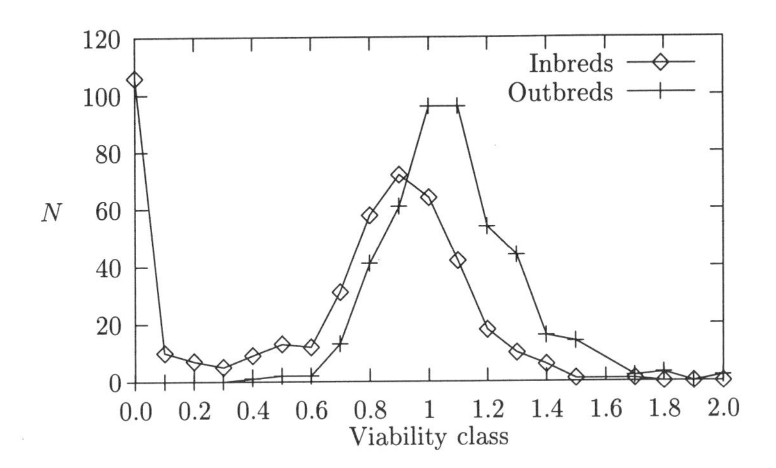

Figure 3.6: The numbers of lines in each viability class among the inbred and outbred flies in the Greenberg and Crow experiment.

The mode of the deleterious peak lies a bit to the left of the mode of the outbred flies. Thus, flies homozygous for their second chromosomes do not fare as well as outbred flies, even if their second chromosomes are free of lethal mutations. Being homozygous at all loci is clearly bad; this is an example of a more general phenomenon called inbreeding depression, which could be defined as the reduction in mean fitness due to increased homozygosity.

At this point, it might seem that we have in hand some evidence on the value of the heterozygous effect, h, for the hidden variation uncovered in this experiment. Perhaps the observation that inbreds are less viable than outbreds implies that overdominance is the prevalent form of dominance. While plausible, inbreeding depression does not allow this conclusion, as may be shown by considering the contribution of a single second-chromosome locus to viability. The fitnesses and frequencies of the genotypes at this locus in inbred and outbred flies of the Greenberg and Crow experiment are as follows:

Genotype:	A_1A_1	A_1A_2	A_2A_2
Relative fitness:	1	$1-hs$	$1-s$
Inbred frequencies:	p	0	q
Outbred frequencies:	p^2	$2pq$	q^2

The experiment tells us that the mean fitness of the inbred flies is less than that of the outbred flies. What does this say about the values of s and h in the table?

The mean viability of the inbred flies is

$$(p \times 1) + [q \times (1-s)] = 1 - qs.$$

The mean viability of the outbred flies is just the mean fitness of a randomly mating population, as given in Equation 3.3. If the outbreds are more viable than the inbreds,

$$1 - 2pqhs - q^2s > 1 - qs \tag{3.12}$$

In sequence, cancel the ones, cancel qs, move q to the right side of the inequality, and clean up to get

$$\boxed{h < 1/2} \tag{3.13}$$

Thus, inbreeding depression implies that h is less than one-half. Recalling that in the s-h parameter system the A_1A_1 genotype is always more fit than A_2A_2, we can conclude that inbreeding depression implies that the fitness of the heterozygote is, on average, closer to that of the more fit homozygote. Recessive deleterious mutations have this property, as do overdominant mutations. While intuition may have suggested that inbreeding depression implies overdominance, we now see that it only limits the heterozygous effect to being less than one-half. Of course, this is a tremendous step forward in our quest to learn more about the alleles responsible for genetic variation in fitness. Already we can pay less attention to underdominant and recessive advantageous mutation and focus on those mutations with $h < 1/2$.

The next step in Greenberg and Crow's analysis is an indirect but brilliant inference about the relationship between h and s, which begins with some formulae that relate the mean relative viabilities of flies to the frequencies and effects of deleterious mutations. Three viability estimates are required:

A The average relative viability of outbred flies. From Figure 3.6 we have $A = 1.008$.

B The average relative viability of inbred flies: $B = 0.632$. When other studies are included, B is found to lie between 0.614 and 0.656.

C The average relative viability of inbred flies without lethal mutations on their second chromosomes: $C = 0.842$. Among all such studies, C ranges from 0.829 to 0.860.

Each second chromosome is imagined to have an unknown number, n, of loci that are capable of mutating to deleterious and lethal alleles. The frequency of the deleterious allele at the ith locus is called q_i and its selection coefficient, which is thought to be small, s_i. Similarly, the frequency of the lethal mutation at the ith locus is called Q_i and its selection coefficient, which is close to one, S_i. Inbred flies have lower viabilities than outbred flies because they are more likely to carry one or more of these mutations in the homozygous state. The probability that a particular inbred fly is homozygous for a deleterious allele at the i locus is q_i. The probability that it dies from this allele, given that it is homozygous, is s_i. The probability that it is both homozygous and dies is q_is_i. Finally, the probability that it survives is $1 - q_is_i$. The situation at a typical locus, the ith locus, is illustrated in Figure 3.7.

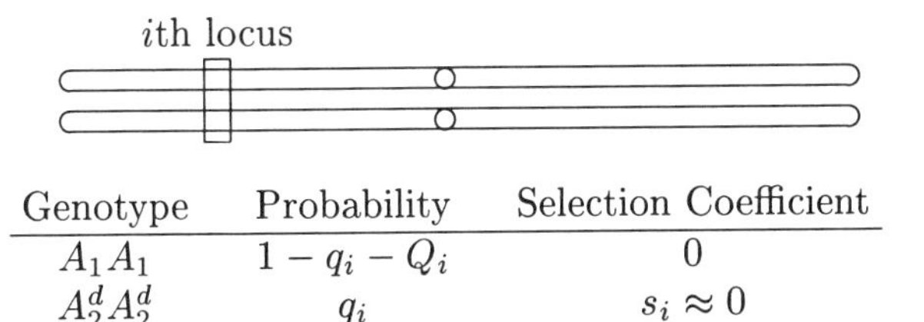

Genotype	Probability	Selection Coefficient
A_1A_1	$1 - q_i - Q_i$	0
$A_2^dA_2^d$	q_i	$s_i \approx 0$
$A_2^lA_2^l$	Q_i	$S_i \approx 1$

Figure 3.7: The possible states of a typical locus in an inbred fly. A superscript d on an allele indicates that it is a deleterious mutant. A superscript l identifies a lethal mutant.

If—and this is a big 'if'—the loci act independently in their effects on the probability of survival, the probability that a particular inbred fly survives to adulthood is

$$B = A\prod_{i=1}^{n}(1 - q_i s_i)(1 - Q_i S_i). \tag{3.14}$$

The factor A represents the probability of survival for an outbred fly. Inbreds may die for all of the various reasons that outbreds may die, plus some more, which are captured in the product term. Using Equation A.4, Equation 3.14 may be approximated by

$$B = Ae^{-D-L}, \tag{3.15}$$

where $D = \sum q_i s_i$ is called the detrimental load and $L = \sum Q_i S_i$ is called the lethal load. These loads differ from the genetic load described in the previous section in that they describe the fitness reduction in inbred rather than outbred flies.

By taking the natural logarithm of both sides of Equation 3.15 and rearranging, we obtain

$$\begin{aligned} D + L &= \ln(A) - \ln(B) \\ &= \ln(1.008) - \ln(0.632) \\ &= 0.4668, \end{aligned} \tag{3.16}$$

which relates the estimates of relative viabilities from Figure 3.6 to the parameters of deleterious alleles. D itself may be obtained from

$$\begin{aligned} C &= A\prod_{i=1}^{n}(1 - q_i s_i) \\ &\approx A\exp(-\sum_i q_i s_i) \\ &= Ae^{-D} \end{aligned}$$

by taking logarithms of both sides and rearranging to obtain

$$D = \ln(A) - \ln(C) = 0.1799.$$

This result may be combined with Equation 3.16 to get

$$L = \ln(C) - \ln(B) = 0.2868.$$

The final quantity of interest, the $D : L$ ratio, is

$$\frac{D}{L} = \frac{\sum q_i s_i}{\sum Q_i S_i} = 0.627. \tag{3.17}$$

This concludes the association of estimated values from the experiment to parameters from the population. The next task is to interpret the result.

In the late 1950s, most population geneticists believed that the majority of mutations in natural populations are deleterious with similar, and small, heterozygous effects. Greenberg and Crow claimed that this view was not compatible with a $D : L$ ratio of 0.627 by the following argument. Using Formula 3.9 for the equilibrium values of q_i and Q_i and canceling the selection coefficients, Equation 3.17 becomes

$$\frac{D}{L} = \frac{\sum u_i/h_i}{\sum U_i/H_i} \approx \frac{\sum u_i}{\sum U_i},$$

where u_i and U_i are the mutation rates to deleterious and lethal mutations at the ith locus, respectively. The hypothesis of similar heterozygous effects, $h \approx h_i \approx H_i$, allows the final cancelation of the heterozygous effects. Under this hypothesis, the $D : L$ ratio is equal to the ratio of the total deleterious mutation rate to the total lethal mutation rate. These rates may be estimated in the laboratory and their ratio, circa 1960, was known to be between 2 and 3. Spontaneous deleterious mutations are two to three times more likely to have a small effect than to be lethal. Thus, the hypothesis of equal heterozygous effects must be rejected because of the small value of the $D : L$ ratio.

Greenberg and Crow suggested an alternative hypothesis: Suppose there is an inverse relationship between h and s. For example, suppose hs is constant across mutations. In this case,

$$\frac{D}{L} = \frac{\sum (u_i/h_i s_i) s_i}{\sum (U_i/H_i S_i) S_i} = \frac{\sum u_i s_i}{\sum U_i S_i},$$

because, by assumption, $h_i s_i \approx H_i S_i$. Now the $D : L$ ratio is the ratio of quantities that are sums of selection coefficients weighted by their mutation rates. These quantities may also be estimated in the laboratory, and the ratio turns out to be 0.711, remarkably close to 0.627, thus giving support to the hypothesis of an inverse homozygous-heterozygous effect.

Greenberg and Crow go on to consider a few other hypotheses but finally return to this one and with it their suggestion that there is an "inverse heterozygous-homozygous effect" for deleterious mutations uncovered by inbreeding.

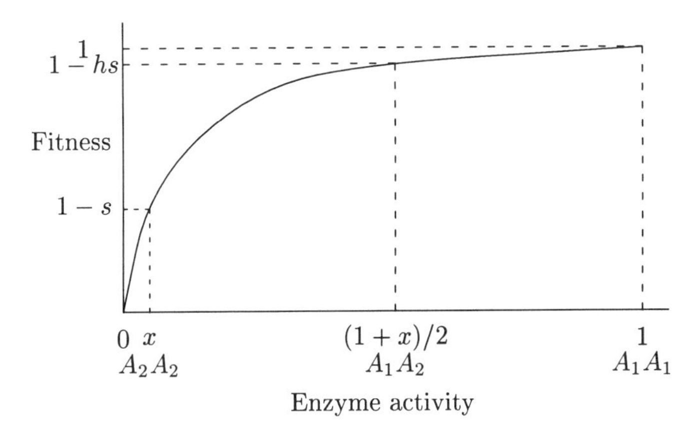

Figure 3.8: Sewall Wright's model of dominance applied to viability.

Mutations of large effect, like lethals, are almost recessive ($h \approx 0$); alleles of small effect have a much greater heterozygous effect, perhaps close to one-half, but not greater than one-half because of inbreeding depression.

The inverse heterozygous-homozygous effect is not an isolated phenomenon. Rather, it is an instance of a more general framework whose origins may be traced to the 1920s and the first efforts to understand the dominance relationships between alleles affecting phenotypes. Some visible phenotypes also exhibit an inverse homozygous-heterozygous effect, the most celebrated being the white-eye mutants in *Drosophila*.

Sewall Wright, among others, created a model to explain the inverse effect that used well-established properties of enzyme pathways (Wright 1929, 1934). The model is of a locus whose product, an enzyme, catalyzes one step in a critical enzyme pathway. The most important property of the model is illustrated in Figure 3.8. The horizontal axis is the activity of the enzyme in units chosen so that an activity of one is "normal." The vertical axis is the fitness of the genotype as a function of its enzyme activity. (In the original formulation the vertical axis was a quantitative measure of the phenotype.) If there is no activity, the pathway does not function and the fitness is zero. As the activity increases, there is a rapid increase in fitness because the pathway can now produce its product. With further increases in activity, the pathway begins to function normally and the augmentation of fitness decreases. There is a "law of diminishing returns" as enzyme activity approaches normal levels. It is fairly well established that most enzymes have such high activities that a reduction in activity of one-half has only a small effect on the functioning of the pathway. The figure is drawn to reflect this.

In Wright's model, a lethal mutation is one that produces a defective enzyme with no activity. In the context of our two-allele model, the A_2A_2 homozygote has no activity and dies. The A_1A_1 homozygote produces normal enzymes and has a scaled activity of one. The activity of the A_1A_2 heterozygote is one-half that of a normal homozygote because it contains one allele making a normal enzyme and one making a defective enzyme. Its scaled activity is 1/2, yet its fitness is close to that of A_1A_1. In this case, we have $s = 1$ and h close to zero.

A deleterious mutation of small effect will be one whose enzyme activity is only slightly less than the normal activity. Again, the heterozygote's activity will be halfway between the normal activity and the mutant homozygote activity. The fitness of the heterozygote, however, will be only slightly less than halfway between that of the two homozygotes because of the concave form of the fitness function. In this case, we have s close to zero and h close to one-half. Hence, the inverse homozygous-heterozygous effect of the Greenberg-Crow experiment.

For example, the function illustrated in Figure 3.8 is

$$w(x) = \frac{(1+a)x}{a+x},$$

with $a = 0.06$. When A_2 is a lethal mutation, we have

$$\begin{aligned} w_{11} &= w(1) = 1 \\ w_{12} &= w(1/2) = 0.946 \\ w_{22} &= w(0) = 0. \end{aligned}$$

From these we get $s = 1$ and $h = 0.053$, which represent a high degree of dominance for the A_1 allele. If A_2 is only slightly deleterious, with an enzyme activity of 0.99, then

$$\begin{aligned} w_{11} &= w(1) = 1 \\ w_{12} &= w(0.995) = 0.9997 \\ w_{22} &= w(0.99) = 0.9994, \end{aligned}$$

which gives $s = 5.7 \times 10^{-4}$ and $h = 0.4976$. The alleles in this case have a very small homozygous effect and are very nearly additive.

Wright's model of dominance gives a biological context for Greenberg and Crow's inverse homozygous-heterozygous effect. Without it, we have a couple of apparently contradictory observations about the dominance relationships between alleles of large and small effects. With it, an appropriate response to the Greenberg and Crow experiment would be: How could it be otherwise? Ironically, the realization that Wright's model is relevant to the inverse homozygous-heterozygous effect has come only recently.

Problem 3.12 *The fitness function in Figure 3.8 was drawn using the function* $(1+a)x/(a+x)$ *to match the results of the Greenberg-Crow experiment. Use this function to find* h *as a function of* s*. Plot this function to illustrate the inverse homozygous-heterozygous effect.*

There is no room in our functional view of dominance for overdominance. Wright's model, in particular, allows only incomplete dominance. The compelling biological appeal of his model suggests that overdominance, when it exists, must be due to unusual circumstances. The sickle-cell polymorphism is an example: Overdominance results from the interaction of the solubility of a globin mutation and the growth requirements of a protozoan. This very peculiar situation would probably not generalize to a sizeable fraction of loci.

The analysis of the Greenberg and Crow experimental results may strike some as unacceptably abstract because it gives no examples of deleterious alleles of small effect. Of course, in 1960 there were no molecular techniques to allow this. Today, such techniques are available and so are some examples of mutations that could plausibly contribute to Greenberg and Crow's detrimental class. Null alleles, alleles at enzyme-encoding loci that lack measurable activity, are the best-known group of putative deleterious mutations. In a large study of null alleles in *Drosophila* populations, Chuck Langley, Bob Voelker, and their colleagues (Voelker et al. 1980; Langley et al. 1981) estimated that the average frequency of null alleles is $\bar{q} = 0.0025$. Previously, Terumi Mukai and Clark Cockerham (Mukai and Cockerham 1977) had estimated the average mutation rate to null alleles to be $\bar{u} = 3.86 \times 10^{-6}$. If the nulls are deleterious and the population is at equilibrium, Equation 3.9 gives

$$\overline{hs} = \bar{u}/\bar{q} = 1.5 \times 10^{-3}.$$

Surprisingly, all but 1 of the the 20 autosomal null alleles showed no obvious deleterious effects when homozygous in the laboratory, so these 19 do not contribute to the lethal class of the Greenberg and Crow experiment. No effort was made to estimate the relative fitnesses of the nulls as homozygotes or heterozygotes, so we cannot say what their heterozygous effects are. These estimates will probably never be made because measuring fitness differences as small as 10^{-3} is impractical, if not impossible.

The main insights of the Greenberg-Crow experiments have withstood the tests of time remarkably well. In a monumental experiment, Terumi Mukai measured viability effects of newly arising mutations and directly confirmed the existence of the inverse homozygous-heterozygous effect (Mukai et al. 1972). Mukai also looked at viability variation in natural populations and found no evidence for overdominance, thus supporting our view that overdominance is not a common phenomenon. The ratio of deleterious to lethal mutation rates is now thought to be closer to 10 than to 2 to 3 as used by Greenberg and Crow. The paper by Mike Simmons and James Crow (1977) contains a good summary of these more recent results.

Problem 3.13 *How much more likely is a* Drosophila *in nature to die from a deleterious mutation rather than a lethal mutation? How much more likely is one to die from being heterozygous for a deleterious mutation than homozygous?*

3.6 Changing environments

Greenberg and Crow and subsequent workers have shown that alleles with very small effects on viability are close to additive. Many of these alleles in natural populations are undoubtedly maintained by mutation-selection balance; others could be maintained by balancing selection. However, our theoretical investigations of natural selection showed that balancing selection occurs only when there is overdominance, $h < 0$. If all of the experimental work, except for a few examples of strong selection like the sickle-cell polymorphism, argues against overdominance, it would be rather silly to suggest that balancing selection is an important contributor to genetic variation in fitness. Or would it? Perhaps there are situations where balancing selection can occur without overdominance.

Two of our examples of selection suggest such a situation, that of selection in a changing environment. In *Panaxia*, the fitness of the *medionigra* allele must have changed from advantageous to disadvantageous at least once in the recent past. In addition to this temporal variation in fitness, there must also be spatial variation in fitness, as *medionigra* was in moderate frequency only in the Oxford population. Similarly, the overdominance of the sickle-cell allele occurs only in areas with high levels of malaria. In other areas, there is incomplete dominance. As instructive as these examples are, the idea that the fitness of a genotype depends on the environment is so obviously true that little in the way of support needs to be mustered. If fitnesses do depend on the state of the environment, as they surely must, then they must just as assuredly change in both time and space, driven by temporal and spatial fluctuations in the environment.

Can changing fitnesses in time and space result in balancing selection without overdominance? The answer is yes, but the route to the answer is one of the more difficult proofs in theoretical population genetics. While the proof is difficult, the result is rather intuitive. Imagine that the fitnesses of the A_1A_1 and A_2A_2 homozygotes can be written as $w_{11} = 1 + s_i$ and $w_{22} = 1 - s_i$, where s_i is a selection coefficient whose value depends on the state of the environment in the ith habitat. Imagine further that the fitness of the heterozygote is one, which corresponds to $h = 1/2$. If the environment changes such that s_i is positive in some habitats and negative in others, then a balanced polymorphism is possible if the absolute value of the mean of s_i across habitats is less than the variance in s_i,

$$|E\{s_i\}| < \text{Var}\{s_i\}.$$

That is, if the variance in fitness across habitats is large enough to overcome any mean advantage one allele may have over another, then a balanced polymorphism will occur without overdominance.

Although the general proof of this result is quite difficult, the core of the argument is actually rather simple and will be illustrated here with a very particular example. The example is of spatial variation in fitness. We imagine a simple pattern where, in some parts of a species' range, the A_1 allele is favored over the A_2 allele, and in other parts the opposite. Were there no migration, it is clear that the A_1 allele would fix in those places where it is favored, the A_2 allele would fix where it is favored, and the species would be polymorphic because of balancing selection without overdominance. We could end here with a satisfied grin except for one awkwardness: real species migrate. Migration tends to make species genetically more uniform across their range and can promote

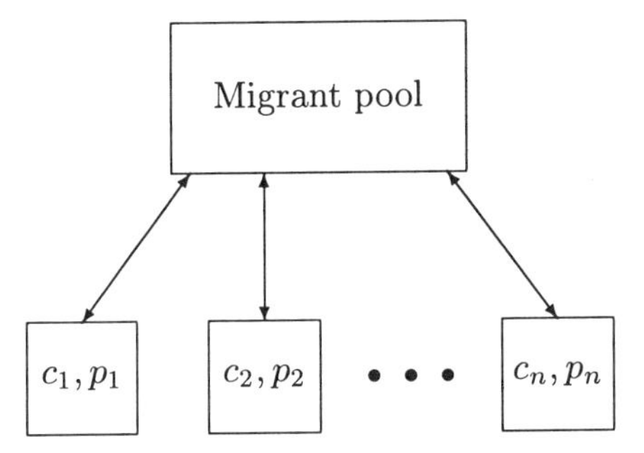

Figure 3.9: A model of selection in a species with subpopulations, each of different relative size, c_i, and fitness.

the fixation of an allele that is more fit on average than all others across the entire range of the species. Obviously, we have some work to do; we must investigate the conditions that lead to polymorphism when migration occurs between subpopulations.

Consider the model of selection in a spatially fluctuating environment illustrated in Figure 3.9. There are n subdivisions in the environment, called patches, each of relative size c_i, $\sum c_i = 1$. In each generation, after selection occurs within a patch, a fraction m of alleles from each patch are exchanged at random with alleles from the other patches. The probability of an exchange with a particular patch is proportional to the relative size of that patch, c_i. The fitnesses of the three genotypes at the A locus vary across patches. The change in p_i, the frequency of the A_1 allele in the ith patch, will be written $\Delta_s p_i$. The frequency of A_1 in the ith patch after a round of selection and migration is

$$p_i' = (1 - m)(p_i + \Delta_s p_i) + m \sum_{j=1}^{n} c_j (p_j + \Delta_s p_j). \tag{3.18}$$

An allele within a patch is from a resident with probability $1 - m$ and from an immigrant with probability m. If it is from an immigrant, then the probability that it is A_1 is just the frequency of A_1 in the entire species after selection. It is as if during each generation all of the subpopulations contribute to a "migrant pool" from which they then choose alleles to replace those that emigrated. Notice that alleles rather than genotypes migrate. We obviously lose some biological realism with this sort of migration, but we gain a great deal in the simplicity of the model. Nothing that we say would change significantly were we to use a model with individuals migrating.

Equation 3.18 cannot be analyzed in the same way as the other models of selection because it is much too complex. However, by examining some special cases we will find out everything we need to know. The first specialization is to one of two extreme migration rates, no migration, $m = 0$, or total migration, $m = 1$; the second is to incomplete dominance, $0 < h < 1$.

For the no-migration case, $m = 0$, the behavior of the model is simple: fixation will occur in each patch for the favored allele. For those patches where

$w_{11} > w_{22}$, $p_i = 1$, otherwise $p_i = 0$. As the frequency of A_1 for the species is the average allele frequency across subpopulations,

$$p = \sum_{i=1}^{n} c_i p_i.$$

The frequency of A_1 is simply the fraction of the entire population for which A_1 is favored and $p_i = 1$. As long as both A_1 and A_2 are each favored in at least one patch, the species will be polymorphic due to balancing selection without overdominance.

The analysis of the total migration case, $m = 1$, begins by examining the change in the species allele frequency,

$$p = \sum_{i=1}^{n} c_i p_i.$$

Multiply both sides of Equation 3.18 by c_i and add the n equations to get

$$\sum_i c_i p_i' = (1-m)\sum_i c_i(p_i + \Delta_s p_i) + m \sum_i c_i \sum_{j=1}^{n} c_j (p_j + \Delta_s p_j)$$

or

$$p' = p + \overline{\Delta_s p},$$

where

$$\overline{\Delta_s p} = \sum_{j=1}^{n} c_j \Delta_s p_j$$

is the average of the changes in p_i across patches. The change in p in a single generation becomes

$$\Delta_s p = \overline{\Delta_s p}. \tag{3.19}$$

On the surface, it looks as if we just made a difficult problem almost trivial. Equation 3.19 says that the change in the species allele frequency is the average of the changes of allele frequencies in the subpopulations. And the migration rate has totally disappeared! However, we can't know $\overline{\Delta_s p}$ without knowing all of the p_i, and they do depend on the migration rate, as seen in Equation 3.18.*

When there is complete migration, $m = 1$, then Equation 3.19 does contain all of the information because the allele frequencies in all of the patches are equal at the start of each round of selection. Using Equation 3.1, Equation 3.19 becomes

$$\Delta_s p = pq \sum_{i=1}^{n} c_i \frac{p(w_{11}^i - w_{12}^i) + q(w_{12}^i - w_{22}^i)}{p^2 w_{11}^i + 2pq w_{12}^i + q^2 w_{22}^i},$$

*In mathematical terms, Equation 3.19 captures just one dimension of an n-dimensional problem.

where the superscript i on the fitnesses indicates that the fitness depends on conditions in the ith patch. Rather than plowing directly into this equation, first specialize to the case of additive alleles ($h = 1/2$) and introduce some symmetry by setting $w_{11}^i = 1 + s_i$, $w_{12}^i = 1$, and $w_{22}^i = 1 - s_i$, giving

$$\Delta_s p = pq \sum_{i=1}^{n} \frac{c_i s_i}{1 + s_i(p - q)}. \tag{3.20}$$

(In deriving this equation, use $p^2 - q^2 = p - q$.)

All that we need from Equation 3.20 is the answer to our question: Does balancing selection occur with incomplete dominance? The easiest route to the answer is to use end-point analysis as we did in the overdominance case. First, does p increase when small? When $p \approx 0$, the sign of Equation 3.20 is determined by

$$\sum_{i=1}^{n} c_i \frac{s_i}{1 - s_i} \approx \sum_{i=1}^{n} c_i (s_i + s_i^2),$$

where the approximation comes from Equation A.7. A_1 will increase when rare ($p \approx 0$) if

$$\sum_{i=1}^{n} c_i (s_i + s_i^2) > 0,$$

which is the same as

$$-\sum_{i=1}^{n} c_i s_i < \sum_{i=1}^{n} c_i s_i^2.$$

The condition that A_2 increases when rare is obtained in the same way and is

$$\sum_{i=1}^{n} c_i s_i < \sum_{i=1}^{n} c_i s_i^2.$$

Both conditions will be met when

$$\left| \sum_{i=1}^{n} c_i s_i \right| < \sum_{i=1}^{n} c_i s_i^2, \tag{3.21}$$

which is a sufficient condition for a polymorphism. (We have used the argument that, if a and b are both positive and if both $a < b$ and $-a < b$, then necessarily $|a| < b$.) If, for example, the absolute value of s_i is the same in each patch and if A_1 is favored in precisely one-half of the patches, then the left side of the polymorphism condition is zero and the right side is greater than zero; hence, polymorphism will occur without overdominance.

As written, Equation 3.21 is not particularly informative. It can be made more so by recognizing that the left side is the absolute value of the average selective advantage of the A_1 allele. If the s_i are viewed as random quantities in a species with a large number of patches, the notation established in Appendix B may be used to write the average as the expectation of s_i, $E\{s_i\}$. The right side may be approximated with the variance in s_i. From Equation B.2, we know that

$$E\{s_i^2\} = \text{Var}\{s_i\} + E\{s_i\}^2.$$

If the mean and the variance of s_i are both very small and of similar orders of magnitude, then the square of the mean will be much smaller than the variance, $E\{s_i\}^2 \ll \text{Var}\{s_i\}$, so

$$E\{s_i^2\} \approx \text{Var}\{s_i^2\}.$$

Now we can rewrite Equation 3.21 in the much more suggestive form,

$$\boxed{|E\{s_i\}| < \text{Var}\{s_i\}} \tag{3.22}$$

If the magnitude of the average selective advantage of an allele is less than the variance in fitness, polymorphism will occur. Said another way, if the variance in fitness is great enough to overcome the average selective advantage of alleles, polymorphism will occur. The more variable the environment, the more polymorphism.

When more complicated models of selection with incomplete dominance in a random environment are examined, the conditions for polymorphism are almost always in a form similar to Equation 3.22. Typically, there is a coefficient on the right hand side that reflects the particular mixture of temporal and spatial components of the fluctuations and the dominance relationships between alleles.

The condition for polymorphism will never be met unless some of the s_i are positive and some negative. This was the only condition required for polymorphism in the no-migration case, $m = 0$. While this condition is necessary for polymorphism when $m = 0$ and $m = 1$, it is sufficient only when $m = 0$. Migration generally makes polymorphism less likely when there is incomplete dominance because migration averages out the very environmental fluctuations that maintain the variation.

Given the ease with which fluctuating environments maintain variation and the fact that temporal fluctuations can cause the fixation of alleles, it is not surprising that the main alternative to the neutral theory as an explanation for molecular evolution and polymorphism is based on selection in a random environment (Gillespie 1991). With fluctuating environments, there is balancing selection in nature, yet in the laboratory the experimentalist will see incomplete dominance. It could be that most of the variation that Crow and Greenberg called deleterious is maintained by fluctuating environments and should be called something else. At the time of this writing, there is no way to know whether the "deleterious" load is due mainly to alleles held in the population by mutation-selection balance or by balancing selection.

Problem 3.14 *Imagine a species that lives in three subpopulations of relative sizes $1/4$, $1/2$, and $1/4$ in which s_i is 0.1, 0, and -0.11, respectively. Will natural selection maintain variation in this species?*

3.7 Selection and drift

Our discussion of directional selection left the impression that the most fit allele eventually reaches a frequency of one. This is true for alleles of moderate frequency but is definitely not true for alleles with only one or a few copies in the population. These alleles are subject to the vagaries of Mendel's law of segregation and to demographic stochasticity. It is easy to see that the fate of a single copy of an allele with, say, a 1 percent advantage is determined mostly by chance. If its frequency should become moderate, then its average selective advantage can overcome the effects of genetic drift.

The interaction of drift and selection is more complex than that of mutation and drift because the strength of selection changes with the frequency of the allele. (Recall the factor pq in $\Delta_s p$ or examine Figure 3.3.) Natural selection becomes a very weak force for rare alleles, as weak as or weaker than genetic drift when only a few copies of the allele are in the population. For example, there is only one copy of a new mutation, so its frequency when it first enters the population is $1/(2N)$. The strength of selection in this case is roughly $1/(2N)$ times s, which is less than the strength of genetic drift. When the frequency of the allele becomes larger, then the strength of selection is determined mainly by the selection coefficient, s. If $s \gg 1/(2N)$, selection will dominate drift for common alleles. Except for cases where $s \approx 1/(2N)$, selection and drift interact only in the dynamics of rare alleles. We will now describe this interaction.

In a finite population, a new advantageous mutation is usually lost because of genetic drift. This surprising result comes from the formula for the probability of ultimate fixation of the A_1 allele given its initial frequency,

$$\pi_1(p) = \frac{1 - e^{-2Nsp}}{1 - e^{-2Ns}}, \tag{3.23}$$

which applies to the case $h = 1/2$. (The derivation of Equation 3.23 will be postponed until Section 3.8.) The subscript on $\pi_1(p)$ is a reminder that the function refers to the fixation probability of the A_1 allele.

The most important application of Equation 3.23 is for the fixation probability of a new mutation, $p = 1/(2N)$, which is

$$\pi_1\left(\frac{1}{2N}\right) = \frac{1 - e^{-s}}{1 - e^{-2Ns}}. \tag{3.24}$$

If s is so small that $e^{-s} \approx 1 - s$ and if $2Ns$ is so large that $e^{-2Ns} \approx 0$, then $\pi_1 \approx s$. Equation 3.23 is for $h = 1/2$, in which case the selection coefficient of the heterozygote is $s/2$. Thus, the fixation probability is twice the selective advantage of the heterozygote. This result holds more generally,

$$\boxed{\pi_1 \approx 2hs} \tag{3.25}$$

While proof of this assertion for arbitrary dominance will not be given, it is in accord with the general observation that selection on rare alleles depends on the fitness of the heterozygote.

Equation 3.25 tells us, for example, that a new mutation with a 1 percent advantage when heterozygous, $hs = 0.01$, has only a 2 percent chance of ultimately fixing in the population. A 1 percent advantage represents rather strong selection. In a very large population, say $N = 10^6$, 1 percent selection will overwhelm drift once the allele is at all common. Yet, 98 percent of such strongly selected mutations are lost. Think of all the great mutations that failed to get by the quagmire of rareness!

There is a much more important implication of our result. Imagine what happens when a species is challenged by a change in its environment. There may be mutations at many loci which, if fixed, will handle the new environment with a new adaptation. Which of these mutations gets there first, as it were, is determined to a great extent by chance. Adaptive evolution is, by its very nature, random. Were we to construct two perfect replicates of Earth, each with exactly the same species and sequences of environments, the course of evolution would be different in the two replicates. Evolution is not repeatable.

It is also possible for a deleterious mutation to fix in the population. The probability of fixation of the A_2 allele, given an initial frequency q, is given by

$$\pi_2(q) = 1 - \pi_1(1-q) = \frac{e^{2Nsq} - 1}{e^{2Ns} - 1}.$$

For a new mutation, $q = 1/(2N)$, and small s we have

$$\pi_2\left(\frac{1}{2N}\right) \approx \frac{s}{e^{2Ns} - 1}. \tag{3.26}$$

In very large populations, large enough that $2Ns \gg 1$, the probability of fixation of a deleterious allele becomes very small. However, when $2Ns$ is close to or less than one, the fixation of deleterious alleles can occur with reasonable probability. An unexpected application of this observation is in molecular evolution.

In Section 2.3, we showed that the rate of substitution of neutral alleles, k, is the mean number of mutations entering the population each generation, $2Nu$, times the probability of fixation of any one of them, $1/(2N)$. For selected alleles, we need only use $\pi_i(1/(2N))$ for the fixation probability to obtain

$$k = 2Nu\pi_i(1/2N). \tag{3.27}$$

For advantageous mutations, Equation 3.25 shows that

$$k = 4Nush.$$

When compared to the rate of substitution of neutral alleles, $k = u$, we see that selected molecular evolution depends on all three parameters of the model, rather than on the mutation rate alone. Many people have used this contrast to argue that selected molecular evolution is not compatible with the molecular clock but is compatible with neutral evolution.

It is also possible that most of molecular evolution involves the fixation of deleterious alleles. In this case, using Equation 3.26 in Equation 3.27, we have

$$k = \frac{2Nus}{e^{2Ns} - 1}. \tag{3.28}$$

At first blush, this seems like a silly idea. How could most of the amino acid substitutions in proteins be deleterious? Certainly, evolution is not lowering the mean fitness of the population! In fact, this idea has a lot to recommend it. For example, it can explain three observations about molecular evolution that the neutral theory cannot explain. The first is the observation that the rate of substitution of amino acids in proteins is lower than the rate of substitution of nucleotides in noncoding regions. If most amino acid mutations are deleterious and most nucleotide substitutions in noncoding regions are neutral, then the amino acid rate should be lower.

The second observation is that there is only a slight generation-time effect in protein evolution but a pronounced generation-time effect in silent (noncoding) evolution. Recall that the generation-time effect is the observation that creatures with shorter generation times evolve faster than do those with longer generation times. The neutral theory predicts a generation-time effect if the mutation rate per generation is fairly similar across species, as it is thought to be. Mice, for example, should exhibit a rate of substitution that is much higher than that of elephants. However, if most mutations are deleterious, then the probability of fixation of these mutations is lower in mice than elephants because the population size of mice is much larger than that of elephants. The consequent lower rate of substitution in mice cancels the generation-time effect.

Problem 3.15 *Graph Equation 3.28 to see if the rate of substitution of deleterious mutations really does decrease with increasing population size.*

The third observation is the narrow range of heterozygosities across groups of organisms that are thought to have very different population sizes. In Section 2.5 we argued that the narrow range could be due to the effective population sizes of species being more similar than their current actual sizes due to fluctuations in their population sizes. Another explanation involves the idea of an effective neutral mutation rate, u_e. If most mutations are deleterious, then only those with selection coefficients close to $1/(2N)$ will attain observable frequencies in natural populations. The rate of mutation to such nearly neutral alleles, which will be lower in larger populations, is called the effective neutral mutation rate. The smaller effective mutation rate in larger populations makes the mutational input, $2Nu_e$, less sensitive to N than for strictly neutral mutations. As a consequence, the variation in heterozygosity should be less across species with different population sizes.

Tomoko Ohta is responsible for the theory that most protein evolution is deleterious while most silent evolution is neutral. Her theory was incorporated into what is now called the neutral theory, even though natural selection is playing a role, albeit a negative one. The common element in both theories is that genetic drift is the force responsible for the substitution of alleles. Motoo Kimura once lamented that the theory was not called the "mutation-random drift theory," as this better represents the forces involved in the substitution of alleles.

3.8 Derivation of the fixation probability

Because much of our understanding of evolution depends on Equation 3.23, it would be remiss not to discuss its derivation. While the derivation is somewhat more technical than others in this book, it does serve as an entrée into more advanced topics in population genetics.

The derivation of Equation 3.23 begins with a decomposition of the fixation probability,

$$\pi(p) = \sum_{\Delta p} \text{Prob}\{\Delta p\}\pi(p + \Delta p).$$

(The subscript on π_1 will be suppressed in this section, as we will discuss only the fixation probability of the A_1 allele.) Imagine a population with p as the initial frequency of the A_1 allele. In the next generation, p will change by a random amount, Δp, whose value reflects the combined action of genetic drift and selection. Suppose you knew with certainty that p changed to a new value p'. Then, for $\pi(p)$ you could use the fixation probability for a population with initial frequency p'. Of course, you will not know what the value of p will be in the next generation. The best you can do is to say that a particular change occurs with some known probability and then average over these changes. This is precisely what the decomposition does. $\text{Prob}\{\Delta p\}$ is the probability of a particular change in p, and $\pi(p + \Delta p)$ is the fixation probability for the new frequency. The sum is over all possible changes in p.

Using the notation of expectation as described in Appendix B, the decomposition may be written as

$$\pi(p) = E_{\Delta p}\{\pi(p + \Delta p)\}, \tag{3.29}$$

which emphasizes that the decomposition really says that the fixation probability in one generation is equal to the expectation of the fixation probability in the next generation (given p in the first generation).

The next step in the derivation uses a Taylor series expansion of the fixation probability in the second generation,

$$\pi(p + \Delta p) \approx \pi(p) + \pi'(p)\Delta p + \frac{1}{2}\pi''(p)(\Delta p)^2,$$

which follows from Equation A.1. Plug this into Equation 3.29 and use the fact that the expectation of a sum is equal to the sum of the expectations (see Equation B.11) to obtain

$$\pi(p) \approx \pi(p) + \pi'(p)E\{\Delta p\} + \frac{1}{2}\pi''(p)E\{(\Delta p)^2\}.$$

Now subtract $\pi(p)$ from both sides to get

$$\frac{1}{2}\pi''(p)E\{(\Delta p)^2\} + \pi'(p)E\{\Delta p\} = 0. \tag{3.30}$$

The two expectations in Equation 3.30 are the mean and (approximately) the variance in the change in p given its current value. We studied these moments in the context of pure drift in Section 2.7. With selection, the mean change in p is not zero, as in the case of pure drift, but rather is approximately

$$E\{\Delta p\} = (s/2)pq = m(p)$$

for the case of additive alleles ($m(p)$ is a commonly used notation for the mean change in p). The expected value of the square of Δp is very nearly equal to the variance in the change in p, because, by Equation B.2,

$$E\{(\Delta p)^2\} = \text{Var}\{\Delta p\} + E\{\Delta p\}^2,$$

and the square of the mean change in p is very small, by assumption. The variance in Δp due to genetic drift was given in Equation 2.16 and is

$$\frac{pq}{2N} = v(p)$$

($v(p)$ is commonly used for variance in Δp). Now, Equation 3.30 may be written as

$$\frac{1}{2}v(p)\pi''(p) + m(p)\pi'(p) = 0. \tag{3.31}$$

Equation 3.31 is a differential equation whose solution, with the proper boundary conditions, is the fixation probability. More exactly, it is a linear second-order differential equation with the two boundary conditions

$$\pi(0) = 0, \; \pi(1) = 1. \tag{3.32}$$

If the initial value of p is zero, fixation is impossible, hence the first boundary condition. If the initial value is $p = 1$, fixation is a certainty, hence the second.

The solution to Equation 3.31 subject to boundary conditions 3.32 is covered in all elementary books on differential equations. Here, we will speed through the solution. Those readers who have trouble with some of the steps should consult a differential equation text.

First, convert Equation 3.31 to a first-order differential equation by defining

$$f(p) = \pi'(p)$$

(and thus having $f' = \pi''$) and multiply both sides of the equation by $2/v(p)$ to obtain

$$f'(p) + \frac{2m(p)}{v(p)}f(p) = 0.$$

This is a first-order linear differential equation and, as such, is much easier to solve than the previous second-order equation. Multiply both sides of the new equation by the integrating factor $\exp(2Nsp)$, and note that the left side is now the derivative of a product ($(uv)' = uv' + u'v$), so

$$\frac{d}{dp}\left[f(p)e^{2Nsp}\right] = 0.$$

When the derivative of a function is zero, the function must be a constant, so

$$f(p)e^{2Nsp} = \text{constant}$$

and

$$f(p) = \pi'(p) = c_1 e^{-2Nsp},$$

where c_1 is a constant whose value will be determined when the boundary conditions are imposed.

Finally, integrate the last equation with respect to p to obtain

$$\begin{aligned}\pi(p) &= c_1 \int_0^p e^{-2Nsx} dx + c_2 \\ &= c_1 \frac{1 - e^{-2Nsp}}{2Ns} + c_2.\end{aligned}$$

The second constant appears because the indefinite integral is only defined up to an additive constant. (If you differentiate both sides of the equation, you recover the previous one for all values of c_2.) The lower limit of the integral on the right side was chosen for convenience.

To satisfy the boundary condition $\pi(0) = 0$, we require $c_2 = 0$. The boundary condition $\pi(1) = 1$ is satisfied when

$$c_1 = \frac{2Ns}{1 - e^{-2Ns}}.$$

Thus, the fixation probability is

$$\pi(p) = \frac{1 - e^{-2Nsp}}{1 - e^{-2Ns}}$$

as we claimed in Equation 3.23.

The conceptually important part of the derivation is the original decomposition, which is called a backward equation. A backward equation relates events at a future time (ultimate fixation in our case) to events at the origin of the process. Backward equations are frequently employed in population genetics to learn about the likelihood of different outcomes of evolution as well as learning the time required to reach these outcomes. The reader might enjoy reading Kimura's original derivation of the fixation probability published in 1962.

The derivation brings out the distinction between mean effects in evolution, as captured in the mean function $m(p)$, and variance effects, as captured in $v(p)$. When setting out to study a new evolutionary model, the first step is usually to describe carefully the mean and variance effects. If the only source of randomness is genetic drift, then $v(p) = pq/(2N)$ is the proper variance function. However, if the model includes random changes in the environment, then a more complicated variance term must be used. The area of mathematics that gives a description of the evolution of the population once $m(p)$ and $v(p)$ are known is called diffusion theory. Warren Ewens' short book, *Population Genetics* (1969), is an excellent introduction to the use of diffusion theory in population genetics.

3.9 Answers to problems

3.1 With the given fitnesses and allele frequency, Equation 3.1 gives

$$\frac{0.1 \times 0.9 \times [0.1 \times -0.05 + 0.9 \times -0.05]}{0.01 \times 0.9 + 0.18 \times 0.95 + 0.81 \times 1} = -0.004545,$$

so in the next generation $p = 0.1 - 0.004545 = 0.095455$.

3.2 First, divide all fitnesses by the fitness of A_1A_1 to obtain the relative fitnesses 1, 0.7333, 0.6667. Next, use the comparisons of the homozygotes' relative fitnesses to get $s = 1 - 0.6667 = 0.3333$. Finally, use the relative fitnesses of A_1A_1 and A_1A_2 to get $hs = 1 - 0.7333 = 0.2666$ and then the computed value of s to get $h = 0.8$.

3.3 If h is positive, then it must be that $1 - hs \geq 0$ or $h \leq 1/s$. If h is negative, there is no upper bound on its magnitude.

3.10 For the overdominance model $w_{\text{max}} = 1 - hs$, so the genetic load is

$$\begin{aligned} L &= \frac{(1 - hs) - [1 - 2pqhs - q^2s]}{1 - hs} \\ &= \frac{q^2s - hs(p^2 + q^2)}{1 - hs}, \end{aligned}$$

which, with the substitution of the equilibrium allele frequency

$$\hat{p} = (h - 1)/(2h - 1)$$

and some nasty manipulations, becomes

$$L = \frac{sh(1 - hs)}{(2h - 1)(1 - hs)}.$$

The mean fitness of the equilibrium population is larger than that of a population which is fixed for the A_1 allele, yet the load in the latter case is zero if the A_2 allele is not included in the description of the homozygous population. This is but one of the idiosyncrasies in load theory that come from the problem of deciding which genotype should determine w_{max}.

3.11 For simplicity, assume that the probability of carrying a lethal mutation at locus, q, is the same for all loci. If there are m lethal-mutable loci in the haploid genome and 20 percent of these are on a single second chromosome, then the probability of a single second chromosome being lethal-free is

$$1 - 0.23 = 0.77 = (1 - q)^{0.2m}.$$

In a diploid genome there are $2m$ lethal-mutable loci; the probability that an entire genome is lethal-free is

$$(1 - q)^{2m} = \left[(1 - q)^{0.2m}\right]^{1/0.1} = 0.77^{10} = 0.073.$$

The probability that a fly from a random mating population carries at least one lethal mutation is $1 - 0.073 = 0.926$.

3.12 First obtain the enzyme activity for A_2A_2, x, as a function of s by solving $1 - s = (1 + a)x/(a + x)$,

$$x = \frac{a(1-s)}{a+s}.$$

For this x, the enzyme activity of A_1A_2 is $(1 + x)/2$, thus,

$$1 - hs = \frac{(1+a)(1+x)}{1+2a+x}.$$

Use for x the function of s and solve for h,

$$h = \frac{a(1+a)}{2a(1+a)+s}.$$

3.13 The ratio of probabilities is the ratio of the genetic loads for deleterious and lethal mutations,

$$\frac{\sum 2u_d}{\sum 2u_l},$$

by Equation 3.10. As the ratio of the total mutation rate to deleterious alleles is now thought to be at least 10 times the lethal mutation rate, a Drosophila is 10 times more likely to die of a mutation of small effect than it is to die from a lethal mutation.

For a particular locus, the ratio of the probability of death as a heterozygote to that of a homozygote is

$$\frac{2pqhs}{q^2s} \approx \frac{2s}{q},$$

which can be quite large for small q.

3.14 The mean value of s_i is

$$E\{s_i\} = 0.25 \times 0.1 + 0.5 \times 0 + 0.25 \times -0.11 = -0.0025,$$

and the mean value of s_i^2 is

$$E\{s_i^2\} = 0.25 \times 0.01 + 0.5 \times 0 + 0.25 \times 0.0121 = 0.005525.$$

As $|-0.0025| < 0.005525$, Equation 3.21 shows that natural selection will hold the locus in a polymorphic state in this species.

Chapter 4

Nonrandom Mating

Most natural populations deviate in some way from the random mating ideal considered thus far. For example, a species whose range exceeds the distance an individual moves in its lifetime cannot possibly mate at random. Departures from random mating can have profound consequences on the evolutionary dynamics of a species. The various ways in which a species might deviate from random mating are diverse, but we will concentrate on only two important departures from random mating, inbreeding and population subdivision. Both may be examined by first describing a generalized form of the Hardy-Weinberg law that includes a new state variable, F, and then by showing the dependency of F on the level of inbreeding or subdivision.

Inbreeding occurs when individuals are more likely to mate with relatives than with randomly chosen individuals. Inbreeding increases the probability that offspring are homozygous and, if practiced by a non-zero fraction of the population, that individuals in the population are homozygous. You will recall that the heterozygous effects of alleles affecting fitness traits are generally less than one-half; thus, this increase in the frequency of homozygotes has a detrimental effect. Inbreeding depression is the phenomenon of lower fitness with higher levels of inbreeding. This phenomenon is quite general: the fitness of most species decreases with increasing homozygosity. For some, the decrease is dramatic, with complete infertility or inviability after only a few generations of brother-sister mating. Inbreeding depression in humans will be described in Section 4.3, followed by a discussion of the role of inbreeding depression in the evolution of selfing in plants.

A species with restricted migration will appear to be inbred because there are more homozygotes than expected under the assumption of random mating, a condition known as Wahlund's effect. Sewall Wright invented a set of measures of departures from Hardy-Weinberg for subdivided populations called F statistics. Here we will consider only the simplest of these, which is called F_{ST}, and show how F_{ST} has been used to investigate the effects of migration on geographic variation in a species.

4.1 Generalized Hardy-Weinberg

An assumption of the Hardy-Weinberg law is violated when populations do not mate at random. We write the genotype frequencies in this case as follows:

Genotype:	A_1A_1	A_1A_2	A_2A_2
Frequency:	$p^2(1-F)+pF$	$2pq(1-F)$	$q^2(1-F)+qF$

The new state variable, F,is used to describe the deviation of the genotype frequencies from the Hardy-Weinberg frequencies. When $F = 0$, we recover the usual Hardy-Weinberg frequencies. If $0 < F \leq 1$, there is an excess of homozygotes compared to the Hardy-Weinberg expectation. When $F < 0$, there is an excess of heterozygotes.

Note that the pair of state variables p and F are always sufficient to describe the genotype frequencies at a single locus with two alleles. At first, this may seem incorrect because two variables should not be able to describe fully the state of a three-variable system. However, although there are three genotype frequencies, x_{11}, x_{12}, and x_{22}, they must add to one. Thus, the dimensionality of the space of genotype frequencies is two rather than three.

Problem 4.1 *Find p and F for a population in which the genotype frequencies of A_1A_1, A_1A_2, and A_2A_2 are* 0.056, 0.288, *and* 0.656, *respectively.*

F has played such an important role in population genetics that it has acquired a myriad of interpretations. Most of these occur when F is between zero and one, as occurs with inbreeding and subdivision. In this instance, F may be interpreted as the probability of homozygosity due to special circumstances, PHSC (pronounced phonetically). For example, the frequency of A_1A_1 individuals in the population or, equivalently, the probability that a randomly chosen individual is A_1A_1 may be obtained by the following argument.

Draw an individual at random from the population. The probability that one of its two alleles at the A locus is A_1 is p. The probability that the other allele is also A_1, given that the first is A_1, is $F + (1 - F)p$. This latter event includes two mutually exclusive components: either the second allele is A_1 because of homozygosity due to special circumstances, which occurs with probability F, or it is not, which occurs with probability $(1 - F)$. If the individual is not homozygous due to special circumstances, the second allele may still be an A_1 allele, as the second allele is, in effect, drawn at random from the population; thus, the probability that it is an A_1 allele is p. We have established that the probability that the second allele is A_1 given that the first is A_1, is $F + (1 - F)p$. The joint probability of A_1 for the first and second alleles in the chosen individual is now seen to be $p[F + (1 - F)p]$, as given above.

A second interpretation of F, when it is between zero and one, is as the correlation of uniting gametes. A full description of the correlation may be found in Appendix B.

Now that we have a general way of describing the state of a population, the next step is to explore how nonrandom mating affects F. We first consider

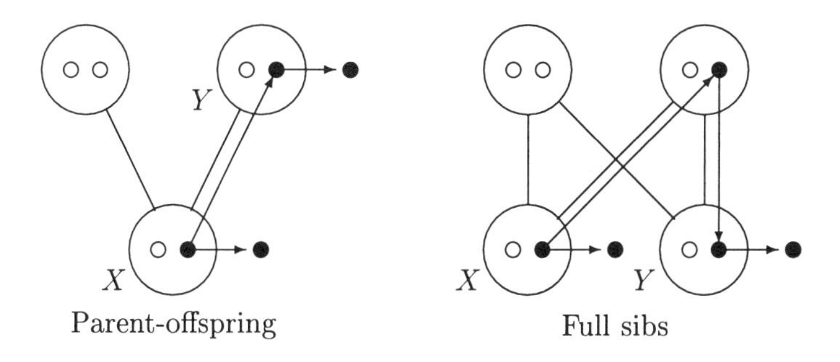

Figure 4.1: Two pedigrees used in the text to illustrate the calculation of the coefficient of kinship.

inbreeding, which occurs when relatives are preferentially chosen as mates. The quantitative aspects of inbreeding depend on the degree of relatedness of mates, which will be explored in the next section.

4.2 Identity by descent

There are many occasions where a quantitative measure of the relationship between relatives is required. In this chapter, we need such a measure to describe inbreeding. In the next chapter, we need it to find the correlation between relatives for quantitative traits. Many different measures are possible; ours, naturally, will be based on the genetic relatedness of relatives. Relatives are genetically similar because they share alleles that are descended from common ancestor alleles. Recall that two alleles at the same locus that are descended from the same ancestral allele somewhere in their recent pasts are said to be identical by descent. Thus, identity by descent is a natural concept to use as the basis of a quantitative description of the relatedness of relatives. One simple measure of relatedness that uses identity by descent is the coefficient of kinship and is usually notated as f_{xy}. The coefficient of kinship is the probability that two alleles, one from individual X and one from individual Y, are identical by descent.

The coefficient of kinship is easy to calculate for simple pedigrees. For complicated pedigrees, one usually uses one of several available algorithms. As all of the pedigrees used in this book and most of population genetics are simple and as the act of reasoning through a pedigree is itself instructive, we will argue from first principles.

The simplest pair of relatives is a parent and its offspring, as illustrated in Figure 4.1. In the figure, the offspring is labeled X and the parent, Y. To find the coefficient of kinship in this case, begin by choosing an allele at random from the offspring and an allele at random from the parent, as illustrated in the figure by the arrows pointing away from alleles in X and Y. The probability that these two alleles are identical by descent is $1/4$, which is obtained in two

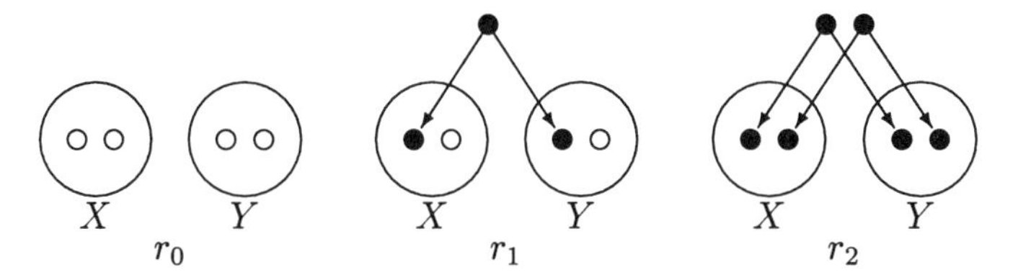

Figure 4.2: The three possible numbers of shared alleles between relatives. The arrows point to pairs of alleles that are identical by descent. r_i is the probability of the particular pattern of shared alleles.

steps. First, the probability that the allele chosen from X came from Y is 1/2. (It is equally likely that this allele came from either parent.) Second, given that the allele came from Y, the probability that it is the one originally chosen from Y is also 1/2. As these two events are independent, the total probability that the two chosen alleles are identical by descent is

$$f_{OP} = \frac{1}{2} \times \frac{1}{2} = \frac{1}{4}.$$

The steps to the coefficient of kinship for full sibs are also illustrated in Figure 4.1. Here one sibling is X and the other is Y. As before, trace backward from the allele chosen from X. This time, the allele will be descended from one of the four alleles in the two parents with certainty. The probability that a copy of the ancestor allele ends up in individual Y is one-half; the probability that that same allele is the chosen allele is also one-half. Thus, $f_{FS} = 1/4$, just as for parent and offspring.

Problem 4.2 *Find the coefficient of kinship for half-sibs and for first cousins.*

The fact that the coefficient of kinship is the same for parent-offspring and full sibs points out the inadequacy of a single number to capture the genetic relatedness of relatives. A parent and its offspring always have exactly one allele each that are identical by descent. (The other allele in the offspring comes from the other parent.) Full sibs, on the other hand, may have zero, one, or two pairs of alleles that are identical by descent. Obviously, another measure of genetic relatedness is needed. The most complete measure is the set of probabilities of sharing zero, one, or two pairs of identical-by-descent alleles, r_0, r_1, and r_2, as illustrated in Figure 4.2. Some examples of these probabilities are given in Table 4.1.

The parent-offspring calculation of r_i should be obvious. The full sib case will take a little more effort. The probability $r_2 = 1/4$ may be reasoned as follows. Pick one of the two alleles in individual X. The probability that an allele identical by descent with this allele is in individual Y is 1/2. Now pick the other allele in X. As it necessarily came from the other parent, its chance of having an identical-by-descent allele in Y is independent of the history of the first allele and is also 1/2. By independence, $r_2 = 1/4$. The converse of this argument gives $r_0 = 1/4$; $r_1 = 1/2$ is obtained by subtraction.

Relationship	r_0	r_1	r_2
Parent-offspring	0	1	0
Full sibs	1/4	1/2	1/4
Half sibs	1/2	1/2	0
First cousins	3/4	1/4	0

Table 4.1: The probability of two relatives sharing zero, one, or two alleles that are identical by descent.

The difference in the two pairs of relatives, parent-offspring and full sibs, is now apparent. A parent and offspring always have one allele each that are identical by descent, while this occurs only one-half the time in full sibs. On the other hand, full sibs may share two or no identical-by-descent alleles, something that never happens with a parent and its offspring. As a consequence, in certain situations full sibs may appear more similar than parent-offspring, but only when there is dominance in the phenotype, as will be shown in Chapter 5.

One needs a certain talent for the calculation of these probabilities to come easily. Personally, I find them difficult and usually pull a book off of my shelf for any but the simplest pedigrees. If they do come easily to you, then by all means enjoy finding the probabilities for other pairs of relatives. For the remainder of this book, only those appearing in Table 4.1 will be required.

The coefficient of kinship may be written in terms of the r_i as

$$\begin{aligned} f_{xy} &= r_0 \times 0 + r_1 \times 1/4 + r_2 \times 1/2 \\ &= \frac{1}{4} r_1 + \frac{1}{2} r_2. \end{aligned}$$

Each term in the sum corresponds to one of three mutually exclusive ways that two chosen alleles might be identical by descent. For example, the second term is the probability that X and Y share exactly one identical by descent allele times the probability that these two alleles are chosen, given that they are shared. The mean number of shared alleles is

$$\begin{aligned} \bar{r} &= 0 \times r_0 + 1 \times r_1 + 2 \times r_2 \\ &= r_1 + 2r_2, \end{aligned}$$

using Equation B.1. One-half the mean number of shared alleles,

$$r = \bar{r}/2 = \frac{1}{2} r_1 + r_2, \qquad (4.1)$$

is called the coefficient of relatedness, r. Some authors use the coefficient of relatedness where we use the coefficient of kinship. The two are related by

$$f_{xy} = \frac{1}{2} r.$$

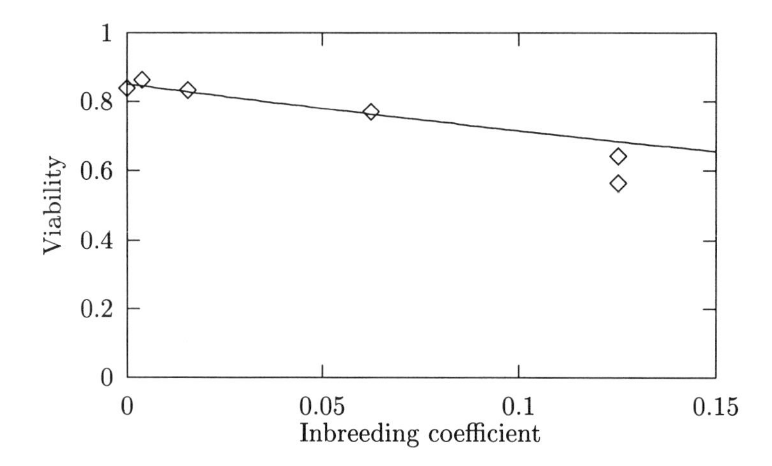

Figure 4.3: The viabilities of young children as a function of their inbreeding coefficients. The curve is the function $\exp(-A - BF_I)$ with $A = 0.1612$ and $B = 1.734$. The data are from Morton et al. (1956).

4.3 Inbreeding

Inbreeding occurs when an individual mates with a relative. The progeny of such matings are more likely to be homozygous than are progeny produced by random matings. The level of inbreeding is measured by the inbreeding coefficient, F_I, which is the probability that the two alleles in an individual are identical by descent. As one of these alleles comes from one parent and the other from the other parent, the inbreeding coefficient of an individual is just the coefficient of kinship of its parents, $F_I = f_{xy}$. For example, the inbreeding coefficient of the offspring of a mating between full sibs is 1/4 because $f_{FS} = 1/4$.

As inbreeding increases the probability of being homozygous and as the average heterozygous effect of alleles affecting viability is less than one-half (at least in *Drosophila*), we would expect that inbred individuals would be less viable than outbred individuals. One demonstration is given in Figure 4.3, which graphs the viability of young human children as a function of their inbreeding coefficient. The functional dependence of viability on F_I may be described theoretically in much the same way that we described the effects of complete inbreeding in the Greenberg and Crow experiment. The particular approach that we will use comes from the classic paper by Newton Morton, James Crow, and H. J. Muller published in 1956.

To derive the mean viability of the offspring of relatives, we require the frequencies of the three genotypes, A_1A_1, A_1A_2, and A_2A_2, among the offspring. First, consider the frequency of the A_1A_1 homozygotes. An offspring could be A_1A_1 for two reasons. Its two alleles could be identical by descent, which occurs with probability F_I, and they both could be A_1, which occurs with probability p (given that they are identical by descent). Thus, the probability of being A_1A_1

for this reason is pF_I. The second way to be an A_1A_1 is to have the two alleles not identical by descent but both A_1 anyway. The probability of this event is $(1-F_I)p^2$. (If the individual does not have two identical-by-descent alleles, then it is as if the individual were the product of random mating.) Similar reasoning leads to the following:

Genotype:	A_1A_1	A_1A_2	A_2A_2
Frequency:	$p^2(1-F_I)+pF_I$	$2pq(1-F_I)$	$q^2(1-F_I)+qF_I$
Fitness:	1	$1-hs$	$1-s$

The mean fitness of the offspring is

$$\begin{aligned}\bar{w} &= 1 - 2pq(1-F_I)hs - [q^2(1-F_I) + qF_I]s \\ &= 1 - a - bF_I,\end{aligned}$$

where

$$\begin{aligned}a &= 2pqsh + q^2 s \\ b &= 2pqs(1/2 - h).\end{aligned}$$

In real populations, $h < 1/2$; therefore, b is always positive. Consequently, $\bar{w}$ is a decreasing linear function of F_I.

The probability of survival of a child, S, is affected by all of its loci plus various nongenetic factors,

$$S = \prod_j (1 - x_j) \prod_i (1 - a_i - b_i F_I),$$

where x_j represents the probability of death due to the jth nongenetic factor and the subscripts on a and b refer to the ith locus. The first product is over all nongenetic factors, and the second product is over all loci. By now, you should have recognized the implicit assumption that the loci act independently. This assumption was also invoked in the analysis of the Greenberg and Crow experiment. With our usual approximation for the product of numbers each of which is close to one, we get

$$S \approx e^{-A-BF_I}, \tag{4.2}$$

where $A = \sum_j x_j + \sum_i a_i$ and $B = \sum_i b_i$.

As the natural logarithm of S is a linear function of F_I,

$$\log(S) = -A - BF_I,$$

it is a simple matter to estimate the values of A and B using standard regression methods. Morton, Crow, and Muller provide the following estimates:

$$A = 0.1612, \; B = 1.734.$$

The curve in Figure 4.3 is plotted using these values of A and B. For small values of F_I,

$$S \approx e^{-A}(1 - BF_I),$$

which explains the roughly linear decrease of viability with F_I for the leftmost points in Figure 4.3.

Problem 4.3 *Suppose all of the inbreeding depression in Figure 4.3 is due to 1000 loci with lethal mutations at mutation-selection equilibrium. If the mutation rate to lethals at each of these loci is 10^{-5}, what are the heterozygous effects of the lethal mutations, h?*

The complete story about the relationship between mean fitness and the inbreeding coefficient is more complicated. The assumption that fitnesses at different loci may be multiplied is at variance with many experimental studies of inbreeding. Generally, the decrease in fitness with F_I goes faster than linear because of synergistic effects of mutations. Even the scant data in Figure 4.3 bear this out as the rightmost two points lie below the curve drawn under the assumption of multiplicative epistasis. As this will be important in the discussion of the evolution of sex in Chapter 6, we will return to the interaction between loci at that time. However, even when synergistic epistasis is present, the decrease in mean fitness with F_I is approximately linear for small F_I, as is the case with the human data.

Inasmuch as inbreeding increases homozygosity and, as a consequence, lowers fitness, it is not surprising that many species have evolved mechanisms to reduce the likelihood of mating with close relatives. Incest avoidance has been documented in many primate societies and, in humans, falls under the rubric of incest taboos, which are present in most cultures.

Now it is time to move from a description of the offspring of matings between relatives to a description of an inbred population. It should come as a pleasant surprise that there is nothing to do other than to note that F in the generalized Hardy-Weinberg described in Section 4.1 is precisely the inbreeding coefficient of the population. In doing so, we imagine a population in which mating with relatives is common. We need not specify the kinds of matings that are taking place, as long as we happen to know F_I. Unfortunately, if we do know the sorts of matings, it is very difficult to find F_I for the population. For example, if we knew that one-half of the time individuals mated at random, one-quarter of the time they mated with their first cousins, one-eighth of the time with their grandparents, and the remaining times equally between full sibs, half sibs, and second cousins, it would be a struggle to determine F_I for the population.

The one example of inbreeding that is easy to study theoretically happens to be the one that is most relevant to natural populations, mixed selfing and outcrossing as practiced in many plant species. Under the mixed model, an individual is created by an act of selfing (literally, mating with one's self) with probability α or by random mating with probability $(1 - \alpha)$. Selfing will, in general, change F_I. The value of F_I in the next generation, as a function of its current value, is

$$F'_I = \alpha[F_I + (1 - F_I)\frac{1}{2}].$$

An individual in the next generation can have two alleles that are identical by descent only if the individual is produced by an act of selfing. Thus, the inbreeding coefficient in the next generation is multiplied by the probability of selfing, α. Given that an individual was produced by selfing, it can have two identical-by-descent alleles for one of two reasons: either its parent had two identical-by-descent alleles, which occurs with probability F_I, or its parent's alleles were not identical by descent, but the two alleles in the offspring are copies of the same allele in the parent, which occurs with probability $(1 - F_I)(1/2)$.

The change in F_I in a single generation is

$$\Delta_\alpha F_I = \frac{2-\alpha}{2}\left(\frac{\alpha}{2-\alpha} - F_I\right).$$

Partial selfing produces two antagonistic forces: selfing increases F_I while outcrossing decreases F_I. Eventually, an equilibrium is reached where $\Delta_\alpha F_I = 0$,

$$\hat{F}_I = \frac{\alpha}{2-\alpha}.$$

Notice that we are able to calculate F_I without making use of allele frequencies; we conclude immediately that allele frequencies do not change under mixed selfing and outcrossing. In fact, allele frequencies do not change under any system of inbreeding. Genotype frequencies do change, and they change in such a way that there are fewer heterozygotes than are seen in an outbreeding population.

Problem 4.4 *Suppose you observed the following genotype frequencies in a plant species that engages in mixed selfing and outcrossing:*

Genotype:	A_1A_1	A_1A_2	A_2A_2
Frequency:	0.828	0.144	0.028

What is the frequency of selfing, α, if the population is at equilibrium?

Problem 4.5 *Graph the inbreeding coefficient as a function of α for an equilibrium mixed selfing and outcrossing population.*

Some interesting evolutionary questions arise with species that are capable of both selfing and outcrossing. For example, in many plant species there is an intrinsic advantage to selfing, which leads to the evolutionary conundrum: Why don't all plant species self? The situation is illustrated in Figure 4.4. The outcrossing pedigree on the right represents a typical individual in an outcrossing population of constant size. This individual leaves behind, on average, two gametes, one carried in an ovule and the other in a pollen grain. These gametes appear as filled circles in the figure.

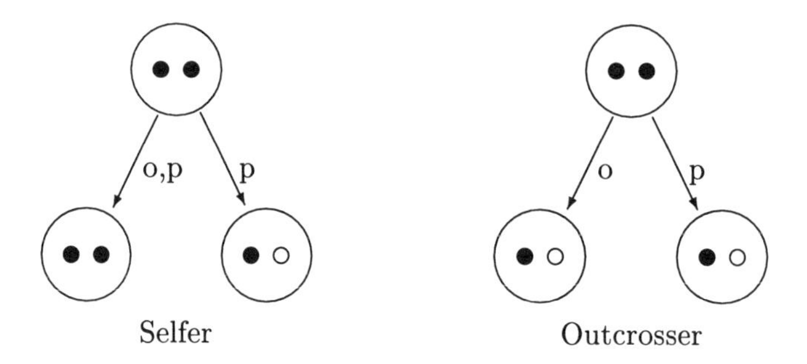

Figure 4.4: The gametes produced by a selfer and an outcrosser. The p to the right of an arrow indicates that the parent's contribution came from pollen; an o indicates it came from an ovule. The filled circles represent gametes from the illustrated parents; the open circles represent gametes chosen at random from the gamete pool.

Suppose a mutant appears that self-fertilizes all of its ovules, as illustrated on the left side of the figure. Suppose also that there is enough pollen in each individual of this species that the few grains needed for self-pollination by the mutant represent a small fraction of the total pollen. As a consequence, the selfing mutant has essentially the same quantity of pollen available for outcrossing as does a nonselfing individual. All else being equal, the selfing mutant will leave behind three gametes for every two of the outcrossing plants, as indicated by the three filled circles in the figure. Two of the gametes are in its selfed offspring; one is in its outcrossed offspring. Thus, the mutant should increase in frequency, perhaps leading to the establishment of selfing as the usual mode of reproduction.

Of course, all else is not equal. The selfed offspring produced by the selfing mutant will be less fit due to inbreeding. The average number of gametes left behind by the selfer is

$$2e^{-A-BF_I} + e^{-A},$$

using Equation 4.2. (Each term in the sum is the number of gametes from the selfer times the fitness of the zygote containing the gametes.) The average number produced by an outbred individual is $2e^{-A}$. The selfing mutant will leave behind more gametes, on average, if

$$2e^{-A-BF_I} + e^{-A} > 2e^{-A},$$

or, because $F_I = 1/2$ for the selfed offspring of an outbred individual,

$$e^{-B/2} > 1/2.$$

Taking the natural logarithm of both sides and rearranging gives the condition for increase of the mutant,

$$B < 2\log(2) = 1.38.$$

Recall that B is a measure of the fitness decrease that occurs with inbreeding. If B is large, the inequality will be violated. Inbreeding depression will be sufficient to drive the selfing mutant from the population. Otherwise, the intrinsic advantage of selfing will be sufficient to override inbreeding depression and the mutant will increase in frequency. For the human data illustrated in Figure 4.3, $B = 1.734$. If we were plants, our inbreeding depression would be too high to allow the evolution of selfing.

This analysis can only tell us about the fate of a new selfing mutant. As the frequency of the mutation increases, a more complicated analysis is required to deal with all of the genotypes that are competing in the population. However, even this simple analysis suggests that plants should be either nearly complete selfers or nearly complete outcrossers. Intermediate levels of selfing will always be at a disadvantage when compared to higher levels of selfing when the inbreeding depression is small. Otherwise, more outcrossing is favored. In fact, plants do tend to be at one extreme or the other. Inbreeding depression is clearly an important contributor to this pattern. Other factors are important as well. For example, weedy colonizing species might self as a way of guaranteeing that a mating occurs when the population density is very low. For a more thorough treatment of the evolution of self-fertilization, read the two papers by Russ Lande and Doug Schemske (Lande and Schemske 1985; Schemske and Lande 1985).

The discussion of the evolution of selfing illustrates a style of inference based on the fate of rare mutants that is commonly used by evolutionists. A trait that is resistant to invasion by all mutants is often called an evolutionary stable strategy, or ESS. John Maynard Smith's book *Evolutionary Genetics* (1989) has many examples of the use of ESS methodology.

Thus far, we have stressed the deleterious effects of inbreeding. However, although the initiation of inbreeding is always detrimental, a population that maintains a steady level of inbreeding is not necessarily worse off than an outbreeding population. Inbreeding increases the effectiveness of selection against partially recessive ($h < 1/2$) deleterious alleles. At equilibrium, the frequency of these alleles will be lower in an inbreeding population than in an outbreeding one and, as a consequence, the genetic load will actually be lower than in the outbreeding population. To see this, we need to repeat the argument leading to $\Delta_s p$ in Equation 3.2, but using the generalized Hardy-Weinberg frequencies. By now you should have little trouble showing that

$$\Delta_s p = (1 - F_I)\frac{pqs[ph + q(1-h)]}{\bar{w}} + F_I \frac{pqs}{\bar{w}},$$

where

$$\bar{w} = 1 - (1 - F_I)(2pqsh + q^2 s) - F_I qs.$$

When the frequency of the deleterious allele is very small,

$$\Delta_s p \approx (1 - F_I)qhs + F_I qs,$$

which is the inbreeding equivalent of Equation 3.8. The equilibrium between the increase of q by mutation ($\Delta_u p \approx -u$ from Equation 3.7) and the decrease by selection occurs when

$$\hat{q}_I \approx \frac{u}{(1 - F_I)hs + F_I s}.$$

Comparing this to Equation 3.9, we see that inbreeding does lead to a decrease in the frequency of the deleterious allele at equilibrium because

$$(1 - F_I)hs + F_I s > hs$$

when $0 < h < 1$, and therefore $\hat{q}_I < \hat{q}$.

The genetic load of the equilibrium population is derived by first approximating the mean fitness by

$$\bar{w} \approx 1 - (1 - F_I)2qsh - F_I qs$$

and then substituting the equilibrium value $\hat{q}$ to obtain

$$\bar{w} \approx 1 - u\frac{2(1 - F_I)h + F_I}{(1 - F_I)h + F_I}.$$

Recall that the genetic load is, by definition,

$$L = \frac{w_{\max} - \bar{w}}{w_{\max}},$$

so that the equilibrium genetic load under selfing is

$$L \approx u\frac{2(1 - F_I)h + F_I}{(1 - F_I)h + F_I}.$$

With complete outbreeding, $F_I = 0$, $L = 2u$ just as we discovered earlier in Equation 3.11. However, when $F_I = 1$, the load is only u, one-half the outbred load. Thus, an equilibrium inbreeding population will have a lower genetic load than an outbreeding population! If, for some reason, an outcrossing population begins selfing, the mean fitness of the population will initially decrease because of inbreeding depression. Then, as selection begins to lower the frequency of deleterious mutations, the mean fitness will increase to a level that is higher than that of the original outbreeding population. This does not mean that the population is better off in a long-term evolutionary perspective. For example, highly inbred populations are less able to generate genetic diversity through recombination. An increase in genetic diversity can both speed up the rate of evolution and aid in the removal of deleterious mutations from chromosomes, as will be shown in Chapter 6.

4.4 Subdivision

Many species occupy such vast geographic areas or have such effective barriers to migration that they cannot behave as a single, randomly mating population. In such cases, there will be genetic differentiation between subpopulations, which leads to departures from Hardy-Weinberg for the entire species. In the case where there is random mating within each subdivision, the genotype frequencies for the entire species are described by a new incarnation of F called F_{ST}. For example, suppose a species is subdivided into two equal-sized patches with $p_1 = 1/4$ in the first patch and $p_2 = 3/4$ in the second. The genotype frequencies in the two patches are the Hardy-Weinberg frequencies given in the first two lines of the table below.

Genotype:	A_1	A_1A_1	A_1A_2	A_2A_2
Frequency in patch 1:	1/4	1/16	3/8	9/16
Frequency in patch 2:	3/4	9/16	3/8	1/16
Frequency in species:	1/2	5/16	3/8	5/16
Hardy-Weinberg frequencies:	1/2	1/4	1/2	1/4

The genotype frequencies for the entire species are the averages of the frequencies in the two patches. For example, the frequency of A_1A_1 is

$$\frac{1}{2} \times \frac{1}{16} + \frac{1}{2} \times \frac{9}{16} = \frac{5}{16},$$

as given in the third line of the table. The frequency of the A_1 allele in the entire species is obviously $p = 1/2$, so the expected Hardy-Weinberg frequencies are as given in the fourth line of the table. Even though each subpopulation has Hardy-Weinberg frequencies, the species as a whole does not. There are too many homozygotes; the excess requires that $F_{ST} = 1/4$, which is obtained by solving

$$2pq(1 - F) = (1 - F)/2 = 3/8.$$

As an observer, you have no way of knowing whether the excess of homozygotes is due to inbreeding, to subdivision, or to some other cause. You could sample the species more carefully until you identify the geographic substructure, but this is often more difficult than it first appears.

This example may be generalized to an arbitrary number of patches with any allele frequency in each. The aim is to find an expression for F_{ST} in terms of the allele frequencies in the subpopulations and the relative sizes of the subpopulations. Later, we will see how to use F_{ST} to investigate the role of migration in the genetic structure of a species.

Let p_i be the frequency of the A_1 allele in the ith subpopulation. Let the relative contribution of this subpopulation to the species or sample be c_i, $\sum c_i = 1$. Let p be the average frequency of the A_1 allele across patches, $p = \sum c_i p_i$, and let $q = 1 - p$. As with the example, the frequencies of genotypes are

Genotype:	A_1A_1	A_1A_2	A_2A_2
In ith patch	p_i^2	$2p_iq_i$	q_i^2
In species:	$\sum c_i p_i^2$	$\sum c_i 2p_i q_i$	$\sum c_i q_i^2$
In species:	$p^2(1 - F_{ST}) + pF_{ST}$	$2pq(1 - F_{ST})$	$q^2(1 - F_{ST}) + qF_{ST}$

By equating the two ways of writing heterozygote frequencies in the species,

$$2pq(1 - F_{ST}) = \sum c_i 2p_i q_i,$$

we get

$$F_{ST} = \frac{2pq - \sum c_i 2p_i q_i}{2pq}.$$

Using

$$2pq = 1 - p^2 - q^2$$

and

$$\sum c_i 2p_i q_i = 1 - \sum c_i p_i^2 - \sum c_i q_i^2 = 1 - \sum c_i (p_i^2 + q_i^2),$$

the expression for F_{ST} may be written as

$$F_{ST} = \frac{\sum c_i (p_i^2 + q_i^2) - p^2 - q^2}{2pq} = \frac{G_S - G_T}{1 - G_T}, \tag{4.3}$$

where

$$\begin{aligned} G_T &= p^2 + q^2 \\ G_S &= \sum c_i (p_i^2 + q_i^2). \end{aligned} \tag{4.4}$$

G_T is the probability that two alleles drawn at random (with replacement) from the entire species are identical by state. G_S is the probability that two alleles drawn at random (with replacement) from a randomly chosen subdivision are identical by state. (With probability c_i the ith patch is chosen. Given this, with probability $p_i^2 + q_i^2$ two randomly drawn alleles are identical by state.) F_{ST} is seen as a measure of the difference between the probability that two alleles drawn from within a subdivision are identical compared to the probability that two alleles drawn at random from the species are identical.

F_{ST} may also be written as a function of the variance of the allele frequencies across patches by substituting

$$\sum c_i p_i^2 - p^2 = \mathrm{Var}\{p_i\}$$

and

$$\sum c_i q_i^2 - q^2 = \mathrm{Var}\{q_i\}$$

into the middle first expression for F_{ST} in Equation 4.3. (See Equation B.2 if the allele frequency variance formulae are not familiar.) This gives

$$\begin{aligned} F_{ST} &= \frac{\mathrm{Var}\{p_i\} + \mathrm{Var}\{q_i\}}{1 - G_T} \\ &= \frac{2\mathrm{Var}\{p_i\}}{1 - G_T}. \end{aligned}$$

The final step follows from $q_i = 1 - p_i$ and Equation B.10. This result shows that F_{ST} is always positive. As long as there is variation in the allele frequency across subdivisions ($\mathrm{Var}\{p_i\} > 0$), the genotype frequencies of the species will exhibit a deficiency of heterozygotes, a condition known as Wahlund's effect.

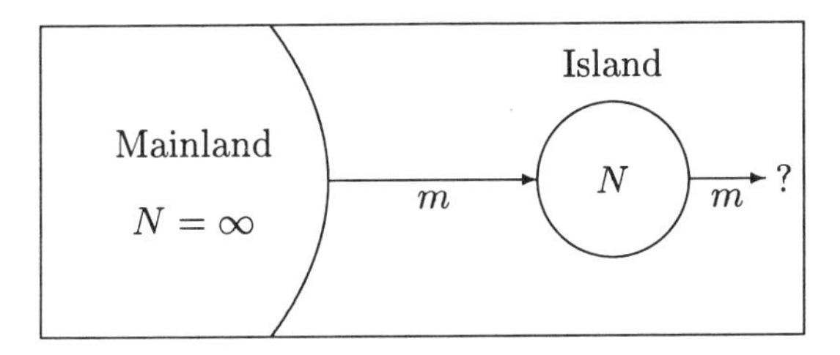

Figure 4.5: The island model.

Problem 4.6 *Verify that the expression for F_{ST} using the variance in allele frequencies gives the correct value for the example at the beginning of this section.*

Inbreeding and subdivision both lead to an apparent deficiency of heterozygotes over Hardy-Weinberg expectations. Thus, nothing can be said about the causes of an observed deficiency without more information. One interesting case where the deficiency of heterozygotes could be attributed to subdivision involves limpets from the intertidal zone of Western Australia as reported in two 1984 papers by Michael Johnson and Robert Black. An electrophoretic survey of protein polymorphism in limpets from the intertidal zone found a deficiency of heterozygotes with F_{ST} values ranging as high as 0.018. A careful study revealed that there was temporal variation in allele frequencies of new recruits. Thus, the limpets at any one spot in the intertidal zone represented a mixture of groups of limpets with different allele frequencies, with a consequent Wahlund's effect.

F_{ST} is a very crude measure of the geographic structure of a species. As it is just a single number, it cannot distinguish between, say, an allele frequency cline from north to south and a hodgepodge of independent allele frequencies in each subdivision. Both patterns produce a positive F_{ST}. There is one very special situation where F_{ST} may tell something about the migration rates of species. This occurs when the segregating alleles are neutral and there are a large number of subdivisions with equal migration rates between them. Sewall Wright was the first to explore this model, which is usually called the island model. As we describe the island model, it will not appear to apply to a large number of subdivisions. Bear with me; we will get to that model after a short development of the literal island model.

The structure of Wright's island model is illustrated in Figure 4.5. There is a very large, effectively infinite, mainland population that sends migrants to the island. There are N diploid individuals living on the island. Each generation a fraction, m, of these individuals depart for places unknown, to be replaced by individuals from the mainland population. For simplicity, we will assume that haploid gametes rather than diploid individuals migrate. While not realistic for many species, this assumption has no effect on the properties of the island model that are of interest. Thus, $2Nm$ gametes migrate each generation.

Genetic drift and migration are the only evolutionary forces acting on the island population. Genetic drift works to eliminate genetic variation on the island, while migration brings in new genetic variation from the mainland. Eventually, an equilibrium is reached with the level of variation on the island determined by N, m, and the amount of variation in the mainland population. There should be a familiar ring to this model. If we were to substitute *mutation* for *migration*, the island model would sound remarkably like the model of mutation and drift described in Chapter 2, which led to Equation 2.6,

$$\hat{\mathcal{G}} = \frac{1}{1+4Nu}.$$

In fact, with an appropriate assumption, the equilibrium level of variation in the island is

$$\hat{\mathcal{G}} = \frac{1}{1+4Nm}. \tag{4.5}$$

The appropriate assumption is simply that each migrant allele is genetically unique. If the mainland population is truly infinite, then this assumption is automatically satisfied as $\hat{\mathcal{G}} = 0$ when $N = \infty$. Thus, the change in $\mathcal{H}$ on the island due to migration is

$$\Delta_m \mathcal{H} = 2m(1-\mathcal{H}),$$

which is the migration equivalent to Equation 2.10 for mutation. The change in $\mathcal{H}$ due to genetic drift is, from Equation 2.2,

$$\Delta_N \mathcal{H} = -\frac{1}{2N}\mathcal{H}.$$

At equilibrium, when $\Delta_m \mathcal{H} + \Delta_N \mathcal{H} = 0$, $\hat{\mathcal{G}}$ is given by Equation 4.5.

The homozygosity, as measured by $\mathcal{G}$, changes fairly quickly from near one to near zero as $4Nm$ changes from being less than one to greater than one. When near zero, genetic drift removes most of the variation from the island. When near one, the island population's homozygosity is like that of the mainland. Thus, when the fraction of migrants is greater than about $1/(4N)$, the effects of isolation become unimportant and the island appears to be infinite in size, as indicated by its value of $\mathcal{G}$. This observation becomes compelling when expressed in numbers of migrants rather than in fraction of migrants. Isolation disappears if $m > 1/4N$, or, equivalently, if $2Nm > 1/2$. The absolute number of diploid migrants each generation is $2Nm$ (strictly, the number of pairs of haploid genomes, as gametes rather than zygotes migrate in our model). If more than one individual migrates every other generation, then the effects of isolation become unimportant. Surprisingly, this statement is independent of the population size of the island. One might have thought that more migration would be required to make large islands like the mainland. However, in large islands drift is a weak force, so less migration is needed to balance drift. The message is clear: very little migration can make a subdivided species appear like one large randomly mating species when neutral alleles are involved.

It is now time to bring F_{ST} back into the discussion and with it a change in the description of the island model. Instead of a mainland-island structure, imagine a large number of patches, each exchanging a fraction, m, of their gametes with all of the other patches each generation, exactly as was assumed in the selection model illustrated in Figure 3.9. Said another way, the $2Nm$ immigrants into each patch each generation are chosen at random from the other patches. If the number of patches increases to infinity, the species population size will be infinite, while the size of each patch will remain fixed at N. From the point of view of a particular patch, the rest of the patches collectively are just like the mainland in Wright's island model.

The homozygosity in each patch is still given by Equation 4.5, which is essentially the same as G_S as defined in Equation 4.4. (The qualification "essentially" refers to an error of approximately $1/2N$ that occurs when the expected value of the homozygosity, G, is equated with the probability of identity by state, $\mathcal{G}$, as discussed in Section 2.7.) G_S is the average value of the homozygosity across subdivisions. Equation 4.3 for F_{ST} also requires G_T, which is the probability that two alleles drawn at random from the species are identical by state. As the species population size is infinite, $G_T = 0$. Thus, for the island model,

$$F_{ST} = \frac{G_S - G_T}{1 - G_T} = G_S = \frac{1}{1 + 4Nm}.$$

If we knew the patch size for a species, N, and its F_{ST}, then we would know its migration rate, assuming that all of the assumptions of the model are met.

This is a big *if*. To use Wright's island model to estimate m or mN, one must be certain that the variation at the locus under study is neutral, that the population is at equilibrium, and that the migration pattern at least approximates that of the island model. Despite the fact that these assumptions are seldom, if ever, met, F_{ST} is frequently used to gain some insights into the genetic structure of a species. For a discussion of the pitfalls of this approach, read the paper by Monty Slatkin and Nick Barton (1989).

Problem 4.7 *Calculate F_{ST} for the data in Table 1.3. Pretend that there are only two alleles by using the S alleles for A_1 and lumping the other alleles into A_2. First, assume that all of the subdivisions are the same size ($c_i = 1/8$). Next, assume that*

$$c_i = \begin{cases} 0.2 & i = 1, 2, 3, 4 \\ 0.05 & i = 5, 6, 7, 8. \end{cases}$$

Does the fact that F_{ST} depends on your assumption about the c_i make you nervous?

4.5 Answers to problems

4.1 The frequency of the A_1 allele is

$$p = 0.056 + \frac{1}{2} \times 0.288 = 0.2.$$

F may be obtained from the frequency of heterozygotes by solving

$$0.288 = 2pq(1 - F) = 0.32(1 - F)$$

to obtain $F = 0.1$.

4.3 Assuming 1000 identical loci with rare lethal mutations,

$$B = \sum_i 2pq(1/2 - h) \approx 2000q(1/2 - h) \approx 2000 \times \frac{10^{-5}}{h}(1/2 - h) = 1.734.$$

Solving the final equality yields $h = 0.0057$.

4.4 The allele frequency is $p = 0.9$ and the inbreeding coefficient is

$$\hat{F}_I = 0.2 = \frac{\alpha}{2 - \alpha}.$$

Solving for alpha yields $\alpha = 1/3$.

4.6 In the example,

$$p = \frac{1}{2} \times \frac{1}{4} + \frac{1}{2} \times \frac{3}{4} = \frac{1}{2}.$$

The variance in p is

$$\text{Var}\{p\} = \frac{1}{2} \times \left(\frac{1}{4}\right)^2 + \frac{1}{2} \times \left(\frac{3}{4}\right)^2 - \left(\frac{1}{2}\right)^2 = \frac{1}{16}.$$

Finally, $G_T = 1/2$. Thus,

$$F_{ST} = \frac{2\text{Var}\{p_i\}}{1 - G_T} = \frac{1}{4},$$

as was given.

4.7 In the first case, $p = 0.7308$, $\text{Var}\{p_i\} = 0.0140$, and $G_T = 0.6066$. Thus, $F_{ST} = 0.0713$. In the second case, $p = 0.7041$, $\text{Var}\{p_i\} = 0.00711$, $G_T = 0.5833$, and $F_{ST} = 0.03416$. It makes me nervous!

Chapter 5

Quantitative Genetics

Quantitative genetics, an area of scientific enquiry closely allied to population genetics, is concerned with the inheritance of quantitative characters, those whose states, like weight, height, or metabolic rate, fall on a quantitative scale. One of the early triumphs of quantitative genetics was R. A. Fisher's demonstration in 1918 that the correlation between relatives may be explained by the combined action of alleles at many loci, each of relatively small effect. Fisher followed in the footsteps of others who, along with Fisher, put to rest the notion that different laws of inheritance applied to discrete and quantitative traits.

The two traditional concerns of quantitative genetics are the correlation between relatives and the response to selection. Both may be investigated by either a purely statistical approach without explicit reference to the states or frequencies of alleles or from a purely genetic approach that does use allele frequencies, Hardy-Weinberg, and other concepts from population genetics. The former is particularly valuable for introducing the main ideas of quantitative genetics with a minimum of effort and will be developed first. Later, the second approach will be used to explore the effects of dominance on the correlation between relatives.

Our goal is to learn enough of the fundamentals of quantitative genetics to be able to discuss some interesting evolutionary questions that involve quantitative traits.

5.1 Correlation between relatives

Figure 5.1 is an example of the phenomenology that stimulated work in quantitative genetics. In this case, the trait is height and the subjects are students from an evolution class at UC Davis and their parents. The left side of the figure is a histogram of the heights of the students in the class. As males and females have different average heights, the data have been adjusted to make the two sexes equivalent. Classically, quantitative genetics is concerned with the variation in height rather than with average height. For example, we might ask: How much of the variation in height is due to genes and how much is due to

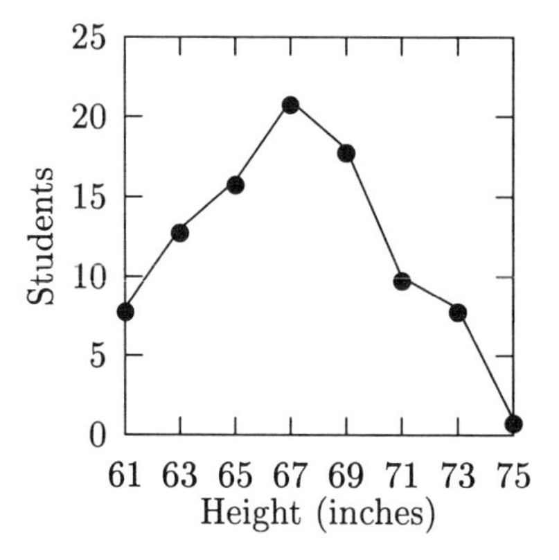

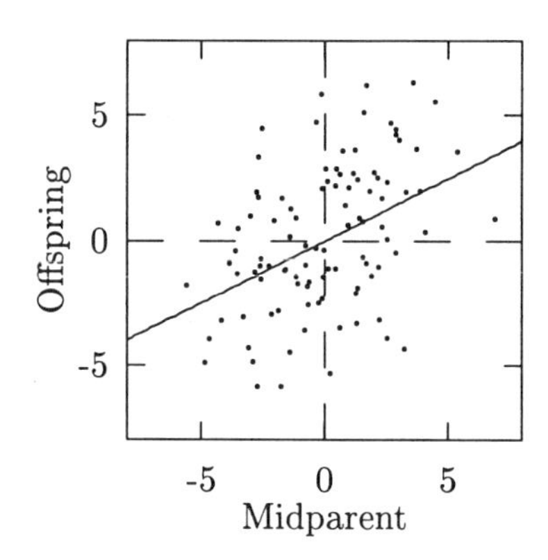

Figure 5.1: The left-hand figure is a histogram of the number of students of a particular height in an evolution class at UC Davis. The right-hand figure graphs the deviation of a student's height from the population mean against the deviation of the student's parents' average height from the population mean.

environmental effects? How much is nature and how much nurture? Quantitative genetics is not, in general, concerned with why the average height of the students is approximately 67 inches.

If genes are important, we might expect relatives to resemble each other. More precisely, we might expect a pair of relatives to look more alike than a pair of randomly chosen individuals. The right side of Figure 5.1 shows that this is true for height in our class. The horizontal axis is the midparent value, which is the average height of the two parents. The values are given as deviations from the mean height. A value of -1, for example, implies that the average value of the two parents was 1 inch below the mean height of the population. The vertical axis is the height of the offspring of the two parents that make up the midparent value. Here, too, the heights are deviations from the population mean. There is a clear relationship between midparent and offspring. There is also a lot of scatter around the regression line due, in part, to Mendelian segregation and, in part, to environmental influences. The fact that large parents tend to have large offspring and small parents tend to have small offspring argues that there is a genetic component to height. But how important is that component? The traditional way to pose the question is to ask: Of the total variation in height, what fraction is attributed to genetic causes and what fraction to environmental causes? The answer is the subject of this section.

To assess the relative contribution of genes to a phenotype, we require an explicit model that states, however naively, the way we imagine a phenotype to be constructed from genetic and environmental factors. The simplest such model, illustrated in Figure 5.2 in the context of two parents and their offspring, posits a single locus that contributes additively to the phenotype and

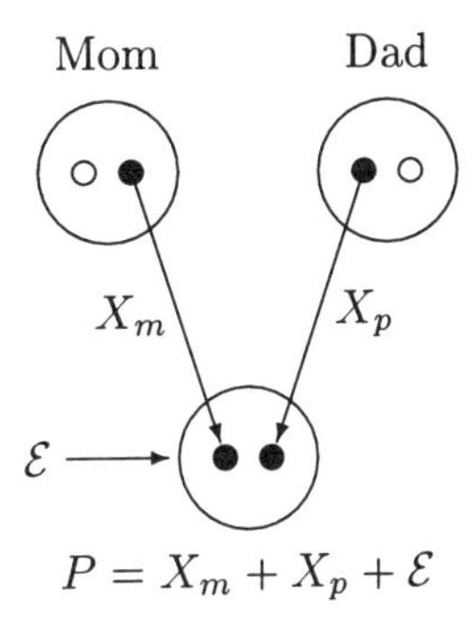

Figure 5.2: The additive model of inheritance for parents and offspring.

an environmental component that also acts additively,

$$P = X_m + X_p + \mathcal{E}. \tag{5.1}$$

P is the value of the phenotype of an individual expressed as the deviation of its value from the population mean. The symbol X_m refers to the additive effect or contribution of the maternally derived allele to the phenotype. For example, this allele might add 1 inch to the height of its carrier. If so, then both the maternal parent and the offspring will be 1 inch taller because of the additive effects of this allele. X_p refers to the paternally derived allele. Its value will, in general, be different from that of the maternally derived allele. The final component is the contribution of the environment, $\mathcal{E}$, which is expressed in the same units as the genetic contribution. If we knew, for example, that $X_m = 2$, $X_p = -1$, and $\mathcal{E} = 5$, then the phenotype would be $P = 6$ units above the population mean.

The mean of P must be zero because all phenotypes are expressed as deviations from the population mean. While this may appear to be a trivial statement, it is worth emphasizing, since we will continually use this fact. The mean values of X_m, X_p, and $\mathcal{E}$ are zero as well. Were they not zero, then the mean value of P would not be zero. Using the notation of expectation, we have

$$\begin{aligned} E\{P\} &= E\{X_m\} + E\{X_p\} + E\{\mathcal{E}\} \\ &= 0 + 0 + 0 = 0. \end{aligned}$$

(If the expectation operator E is not familiar, read about it on page 156.)

Of course, we cannot know the actual values of X_m, X_p, and $\mathcal{E}$ for any particular individual. The best we can do is to view these quantities as random variables and estimate their moments in the population. To be precise, we will assume from henceforth that X_m, X_p, and $\mathcal{E}$ are normally distributed random variables with means equal to zero and variances

$$\mathrm{Var}\{X_m\} = \mathrm{Var}\{X_p\} = V_A/2$$

and

$$\text{Var}\{\mathcal{E}\} = V_E.$$

(The peculiar notation for $\text{Var}\{X_m\}$ and the justification for the normality assumption will be explained shortly.)

The model captured in Equation 5.1 appears to differ fundamentally from those developed in earlier chapters in that alleles are described by their effects on the phenotype. Earlier, alleles were described as genetic entities with frequencies, but often unspecified phenotypes. In fact, the new model is really the same as the old. Imagine a locus with n alleles at frequencies $p_1, p_2, \ldots, p_n$ that contribute $x_1, x_2, \ldots, x_n$ to the phenotype. The number of alleles is imagined to be so large, and their frequencies so small, that the probability of drawing the same allele twice in a finite sample is small. The mean allelic contribution of this locus is

$$\mu = \sum_{i=1}^{n} x_i p_i,$$

and the variance of the contribution is

$$\sigma^2 = \sum_{i=1}^{n} (x_i - \mu)^2 p_i.$$

As the mean effect of a locus must be zero, $\mu = 0$. The notation for the variance of the the allelic contribution introduced above dictates that $\sigma^2 = V_A/2$. The normality assumption implies that if we were to collect all of the x_i and draw a histogram of their values, that histogram would appear to be a bell-shaped curve. The representation of X in terms of allele frequencies and the x_i emphasizes that X is a population quantity rather than an individual or gene-action quantity. x_i, on the other hand, is the phenotypic contribution of a particular allele and is not a population quantity. A common conceptual error is to regard the additive component as being solely a function of gene action, whereas it is really a function of both gene action as captured in the x_i and the state of the population as captured in the p_i.

The reader may well balk at the suggestion that there are so many alleles at a locus and that their phenotypic contributions are normally distributed. We are making this assumption in order to develop quantitative genetics in the simplest context possible. In the last section of this chapter, we will show how the one-locus model may be replaced with a more realistic multilocus model, where each locus may have only a couple of alleles. The normality then comes, when the number of loci is large, from the Central Limit Theorem.

To take the first steps in the analysis of this model, we need to use variances, covariances, and correlations. If you are not familiar with these moments, this is the time to either read the brief summary beginning on page 160 or consult a statistics book (or, better yet, to do both). Quantitative genetics is all about variances, covariances, and correlations. If these ideas are not second nature to you, then it is impossible to understand quantitative genetics!

Equation 5.1 may be used to make precise the question about the relative roles of nature and nurture. As our interest is in explaining variation in quantitative traits, it is quite natural to begin by calculating the variance of both sides of Equation 5.1 (with the help of Equation B.12),

$$\begin{aligned}\text{Var}\{P\} &= \text{Var}\{X_m\} + \text{Var}\{X_p\} + \text{Var}\{\mathcal{E}\} \\ &\quad + 2\text{Cov}\{X_m, X_p\} + 2\text{Cov}\{X_m, \mathcal{E}\} + 2\text{Cov}\{X_p, \mathcal{E}\}. \qquad (5.2)\end{aligned}$$

The variance in the phenotype is the sum of the variances of the genetic and environmental factors and twice the covariances of these factors. The covariance terms will all be set equal to zero. The term $\text{Cov}\{X_m, X_p\}$ is the covariance between the maternal and paternal genetic contributions to the phenotype. As the parents are not related (by assumption), their genetic contributions are independent, which implies that their covariance is zero. $\text{Cov}\{X_m, \mathcal{E}\}$ and $\text{Cov}\{X_p, \mathcal{E}\}$ are covariances between the genetic and environmental contributions and will be assumed to be zero. Were they not zero, we would say that there are genotype-environment interactions. For example, if one allele added one inch to the phenotype in a warm environment and subtracted one inch in a cold environment and another allele did exactly the opposite, there would be a genotype-environment interaction. Genotype-environment interactions are also called norms of reaction. Our only reason for assuming away genotype-environment interactions is to make the model simpler. These interactions do exist in nature but can be minimized in the laboratory.

Without the covariance terms, Equation 5.2 may be written in the much more pleasing form

$$V_P = V_A + V_E, \qquad (5.3)$$

where

$$V_P = \text{Var}\{P\}$$

is the phenotypic variance,

$$V_A = \text{Var}\{X_m\} + \text{Var}\{X_p\}$$

is the additive variance, and

$$V_E = \text{Var}\{\mathcal{E}\}$$

is the environmental variance. The additive variance can also be written as

$$V_A = 2\text{Var}\{X_p\} = 2\text{Var}\{X_m\}$$

because the two parental gametes are chosen at random from the population and, for this reason, are statistically equivalent. The "additive" of additive variance refers to the fact that the genetic contribution is a simple sum of the contribution from each allele. In more complicated situations, the two alleles might interact to produce an additional genetic contribution whose variance is called the dominance variance (described in Section 5.4).

The fraction of the phenotypic variance due to additive effects is simply

$$h^2 = \frac{V_A}{V_P} = \frac{V_A}{V_A + V_E}, \tag{5.4}$$

and is called the heritability of the trait.* For example, if $h^2 = 1/2$, then one-half of the variance in the phenotype is genetic in origin and one-half is environmental. The heritability is precisely the quantity that answers the nature-nurture question. Its value is estimated by using the correlation between relatives, as we will now show.

We begin with an example of a particularly simple pair of relatives: a single parent, mom in this case, and her offspring. The phenotypes of the parent and offspring may be written as follows:

$$\begin{aligned} P_P &= X_m + X'_m + \mathcal{E}_P \\ P_O &= X_m + X_p + \mathcal{E}_O, \end{aligned}$$

where the subscripts P and O refer to parent and offspring. The symbol X'_m represents the genetic contribution to the parent's phenotype from the allele that is not passed on to its offspring (Mom's open circle in Figure 5.2). With this formulation, the resemblance between parent and offspring is seen to come from the shared allele with phenotypic effect X_m, as X_m is the only common factor on the right sides of both equations. The other alleles in each, X'_m and X_p, are no more alike in phenotypic effect than are two alleles drawn at random from the population.

The resemblance between parent and offspring is expressed quantitatively by the covariance of their phenotypes, $\text{Cov}\{P_P, P_O\}$, which is

$$\begin{aligned} &\text{Cov}\{X_m + X'_m + \mathcal{E}_P, X_m + X_p + \mathcal{E}_O\} = \\ &\quad \text{Cov}\{X_m, X_m\} + \text{Cov}\{X_m, X_p\} + \text{Cov}\{X'_m, X_m\} + \text{Cov}\{X'_m, X_p\} \\ &\quad + \text{Cov}\{X_m, \mathcal{E}_O\} + \text{Cov}\{X'_m, \mathcal{E}_O\} + \text{Cov}\{X_m, \mathcal{E}_P\} + \text{Cov}\{X_p, \mathcal{E}_P\} \\ &\quad + \text{Cov}\{\mathcal{E}_P, \mathcal{E}_O\}. \end{aligned} \tag{5.5}$$

The fact that the covariance of a sum is the sum of the covariances comes from Equation B.14. Although the covariance looks horrendous, it is important to appreciate that conceptually it is quite simple. The covariance between parent and offspring must ultimately be due to the covariances of the additive components of each phenotype. There must necessarily be a lot of covariance terms because there are three random quantities contributing to each phenotype. Fortunately, most of these covariances are zero, either by assumption or because of independence, as we will now show.

The first line of covariances on the right side of Equation 5.5 are between the genetic contributions to the phenotypes. The first term in this line is the covariance of a random variable with itself, which is its variance,

$$\text{Cov}\{X_m, X_m\} = \text{Var}\{X_m\} = V_A/2.$$

*Unfortunately, the symbol for heritability is h^2, which has nothing to do with the heterozygous effects of alleles.

The next three covariances are between independently derived alleles. As these alleles are unrelated, the covariances of their effects are zero.

The next line has covariances between a genetic effect and an environmental effect; they are zero because of our previous assumption of no genotype-environment interactions.

Finally, we come to the thorny $\text{Cov}\{\mathcal{E}_P, \mathcal{E}_O\}$, which is the covariance of the environmental effects of parent and offspring. This, too, will be set equal to zero, but not without some misgivings. Often, the environments of parents and offspring are more similar than are those between randomly chosen individuals. This is certainly true for humans, where geography, socio-economic class, level of education, and a myriad of other causes of common environments occur. Common environments will cause relatives to resemble each other more than they would otherwise. As a consequence, common environments may give an experimenter an inflated view of the role of genetics in the determination of the trait. In laboratory situations, the common environment can be minimized and, as most quantitative genetics experiments are done in the laboratory, will be set to zero here.

The expression for the covariance of parent and offspring is now the remarkably simple

$$\text{Cov}\{P_P, P_O\} = \frac{V_A}{2}. \tag{5.6}$$

The covariance between parent and offspring is one-half the additive variance. The presence of the additive variance is traceable to the fact that both the parent and offspring share the allele with value X_m, whose variance is $V_A/2$.

The correlation coefficient between a parent and its offspring is obtained by dividing both sides of Equation 5.6 by V_P and using the definition of the correlation coefficient given on page 161,

$$\begin{aligned}\text{Corr}\{P_P, P_O\} &= \frac{\text{Cov}\{P_P, P_O\}}{\sqrt{\text{Var}\{P_P\}\text{Var}\{P_O\}}} \\ &= \frac{V_A}{2V_P} = \frac{h^2}{2}.\end{aligned}$$

(In the second line we used the fact that the phenotypic variances of any set of unrelated individuals are the same. Thus, the phenotypic variance of parents is the same as that of offspring.) The correlation is an increasing function of the additive variance. For a fixed V_E, the larger the additive variance the greater the resemblance between relatives. Without genetic diversity, relatives would look no more alike than random pairs of individuals. With genetic diversity, they resemble each other because their shared alleles are more similar than random pairs of alleles.

Problem 5.1 *What is the heritability of a trait with $V_A = 2$ and $V_E = 3$? What is the correlation between parent and offspring for this trait?*

The next task is to find the correlation between an arbitrary pair of relatives, X and Y, whose phenotypes are determined by the two equations

$$P_X = X_m + X_p + \mathcal{E}_X$$
$$P_Y = Y_m + Y_p + \mathcal{E}_Y.$$

If we charge ahead, ignoring all genotype-environment interaction and common environment terms, we obtain

$$\text{Cov}\{P_X, P_Y\} = \text{Cov}\{X_m, Y_m\} + \text{Cov}\{X_m, Y_p\} + \text{Cov}\{X_p, Y_m\} + \text{Cov}\{X_p, Y_p\}. \quad (5.7)$$

The values of the covariances depend on the average number of shared alleles between the relatives. Figure 4.2 indicates the possibilities. With probability r_0, the two relatives share no identical-by-descent alleles and all of the covariances in Equation 5.7 are zero. With probability r_1, the two relatives share one pair of identical-by-descent alleles, in which case only the covariance corresponding to the shared allele will be non-zero, and its value is $V_A/2$. (Note that this case is like parent and offspring, for which $r_1 = 1$.) Finally, the relatives could share two identical-by-descent alleles. In this case each pair of identical-by-descent alleles yields one non-zero covariance of magnitude $V_A/2$, so the full covariance when there are two pairs of identical-by-descent alleles is V_A. Considering all three contingencies, the covariance between relatives X and Y is

$$\text{Cov}(P_X, P_Y) = r_0 \times 0 + r_1 \times \frac{V_A}{2} + r_2 \times V_A$$

or, using the coefficient of relatedness defined on page 89, $r = (r_1/2) + r_2$,

$$\text{Cov}(P_X, P_Y) = rV_A. \quad (5.8)$$

The correlation between the relatives is obtained by dividing both sides of Equation 5.8 by the phenotypic variance,

$$\boxed{\text{Corr}\{P_X, P_Y\} = rh^2} \quad (5.9)$$

In words, the correlation between a pair of relatives is equal to the coefficient of relatedness times the heritability. This is certainly one of the most satisfying of answers to an ostensibly difficult problem in all of biology. In his 1918 paper, R. A. Fisher derived the correlation between relatives as a series of special cases for many representative pairs of relatives. The general form of the correlation, of which Equation 5.9 is a special case, was derived in the 1940s by Gustave Malécot and Charles Cotterman.

Problem 5.2 *What is the correlation between each of the pairs of relatives in Table 4.1 if $V_A = 2$ and $V_E = 3$.*

Species	Character	Heritability
Honeybee	oxygen consumption	0.15
Eurytemora herdmani	length	0.12
Cricket	wing length	0.74
Flour beetle	fecundity	0.36
Red-backed salamander	vertebral count	0.61
Darwin's finch	weight	0.91
Darwin's finch	bill length	0.85

Table 5.1: Heritability estimates determined by parent-offspring correlations for a variety of traits and species taken from a paper by Mousseau and Roff (1987).

Heritabilities are easy things to measure. As a consequence, the literature is full of heritability estimates for almost any trait you can imagine. Table 5.1 gives a sample of heritability estimates from an interesting paper by Timothy Mousseau and Derek Roff (1987). Two important generalities emerge from heritability studies. The first is that almost all traits have heritabilities between 0.1 and 0.9. In other words, between 10 and 90 percent of the phenotypic variation seen in most quantitative traits is genetic in origin, an observation that echoes the ubiquitous variation seen in DNA and proteins and poses yet again the question of why so much genetic variation exists.

The second generalization is that life history traits, like viability, longevity, and fecundity, tend to have lower heritabilities (average $h^2 = 0.12$) than do behavioral traits (0.17), which tend to have lower heritabilities than morphological traits (0.32). The reason for this pattern is obscure. Life history traits could have lower heritabilities because of lower additive variances or because of higher environmental variances. The folklore of our field claims that life history traits have lower additive variances because they are closely tied to fitness and that natural selection quickly and efficiently removes additive variation in fitness. However, Dave Houle (1992) recently showed that life history traits actually have higher additive variances than other traits. Apparently, life history traits have lower heritabilities because the joint contributions of environmental, dominance, and epistatic variances are much larger than those in other traits. So much for folklore!

Closely related to our discussion of the correlation between relatives is the answer to the following question: If the phenotype of relative X is $P_X = x$, then what is the expected phenotype of relative Y? For example, if I knew that the average height of a student's parents was 72 inches, what would be the expected height of the student? The answer comes immediately from the regression coefficient as described on page 165. The regression coefficient is the slope of the "best fit" line illustrated in Figure 5.1 and again in Figure 5.3. The line, which is forced to pass through the origin, is fit to the data by the method of least squares, which minimizes the squared deviation (measured vertically)

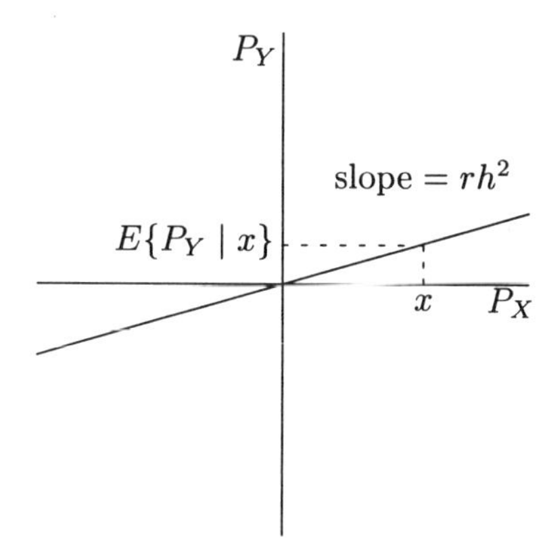

Figure 5.3: The use of regression to find the expected value of the phenotype of relative Y given that the phenotype of relative X is x.

of each point from the line. The slope of the regression line is

$$\beta = \frac{\text{Cov}\{P_X, P_Y\}}{\text{Var}\{P_X\}} = rh^2,$$

using the definition of the regression coefficient in Equation B.17 and Equation 5.8. Thus, the regression of one relative on another is the same as the correlation coefficient of the relatives.

Equation B.18 shows immediately that the expected value of relative Y, given that the phenotype of relative X is x, is

$$E\{P_Y \mid P_X = x\} = \beta x = rh^2 x,$$

where, as usual, all measurements are expressed as deviations from the mean.* Figure 5.3 summarizes the relationship of the mean of P_Y given $P_X = x$. Note that the above relationship holds only if the trait is normally distributed.

The most important implication of this result is that the Y relative is always closer to the population mean, on average, than is the X relative because both r and h^2 are less than one (in most situations). The movement of the mean of relative Y toward the population mean is called the regression toward the mean. It occurs because a pair of relatives share only a portion of their alleles. The rest are obtained at random from the population. Thus, the mean of one relative, given the value of the other, must move toward the population mean.

Finally, we can return to the data of Figure 5.1 to answer the question: What is the heritability of height? One of the "relatives" in the figure is a midparent, which is the average of the phenotypes of the two parents of each student. Consequently, we cannot use the results developed thus far without

*The notation '|' stands for "given" and refers to the assumption that the value of P_X is known.

	Parent-offspring	Midparent-offspring	General
Covariance	$V_A/2$	$V_A/2$	rV_A
Correlation	$h^2/2$	$h^2/\sqrt{2}$	rh^2
Regression coefficient	$h^2/2$	h^2	rh^2

Table 5.2: A summary of the measures of resemblance between pairs of relatives.

some minor modifications. The covariance of midparent and offspring is

$$\begin{aligned}\text{Cov}\{(P_P + P_{P'})/2, P_O\} &= \text{Cov}\{P_P, P_O\}/2 + \text{Cov}\{P_{P'}, P_O\}/2 \\ &= V_A/2,\end{aligned}$$

just as it is for one parent and offspring. However, the variance of midparent is smaller than that of a single parent,

$$\text{Var}\{(P_P + P_{P'})/2\} = V_P/2.$$

By Equation B.8, the correlation of midparent and offspring is

$$\frac{V_A/2}{\sqrt{V_P V_P/2}} = \frac{1}{\sqrt{2}}\frac{V_A}{V_P} = \frac{1}{\sqrt{2}}h^2.$$

Similarly, the regression of offspring on midparent is

$$\frac{\text{Cov}\{(P_P + P_{P'})/2, P_O\}}{\text{Var}\{(P_P + P_{P'})/2\}} = \frac{V_A/2}{V_P/2} = h^2, \tag{5.10}$$

which is not the same as the correlation coefficient for normal pairs of relatives. Table 5.2 is a handy summary of the various measures of relationship between midparent and offspring and those of other pairs of relatives.

Problem 5.3 *The correlation coefficient for the data of Figure 5.1 is 0.476. What is the heritability of height? If one parent's height were 3 inches above the population mean and the other's were 1 inch above, what would be the expected deviation of the height of their child from the population mean?*

Problem 5.4 *The following measurements of weights are from a single parent and its offspring and are expressed as deviations from the mean. What is the heritability of this trait?*

$(-0.002, -0.391)$ $(1.566, 1.747)$ $(0.542, -2.127)$ $(-0.285, -1.623)$
$(-1.519, -0.876)$ $(-1.136, -0.705)$ $(0.907, 0.458)$ $(0.435, -0.287)$
$(1.292, 0.153)$ $(-0.640, 0.711)$

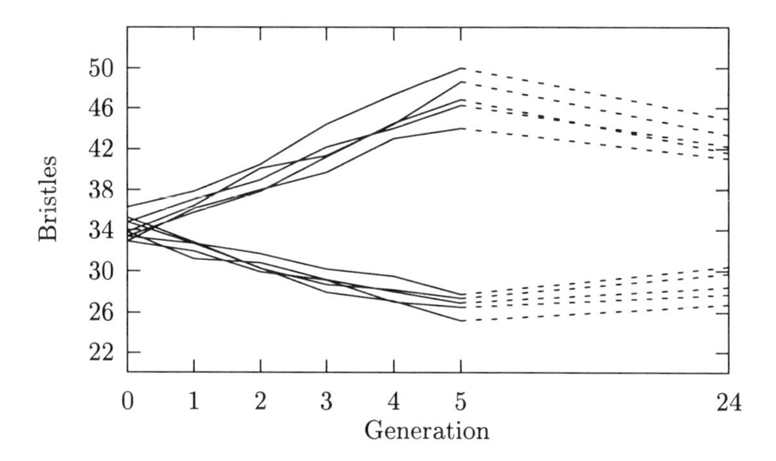

Figure 5.4: The results of a selective breeding experiment for abdominal bristles in *Drosophila*. The upper lines give the number of bristles during five generations of selection for a greater number of bristles in five replicate lines. From generations 6 to 24 there was no selection. The five lower lines give the results of selection for fewer bristles. The data are from Clayton et al. (1957).

5.2 Response to selection

Agriculturalists, from prehistoric times until the present, have improved their crops and livestock by selective breeding: simply choose the best individuals as parents of the next generation. And it works, providing that there is additive variance for the traits of interest. But how well does selective breeding work? Will there be a significant change in the mean value of a trait in a few generations, or are hundreds or even thousands of generations of selective breeding required? The answer is the main goal of this section.

Natural selection on quantitative traits is like selective breeding in that more adaptive morphologies are more likely to leave behind offspring than are less adaptive morphologies. In Chapter 3 we saw how fitness differences between genotypes change allele frequencies. Here, we will see how fitness differences between morphologies change the distribution of morphologies. Surprisingly, we can do this, at least approximately, without explicitly describing changes in genotype frequencies.

Figure 5.4 illustrates a typical selective breeding experiment. In this case the character is the number of bristles on the fourth and fifth abdominal sternites in *Drosophila melanogaster*. Bristle number in *Drosophila* has been studied for years simply because it is easier to count accurately the number of bristles under a microscope than it is to determine more universal quantitative traits like weight or length. The graph shows the mean number of bristles in lines selected for high or low numbers of bristles for five generations. Five replicates lines (the upper five), were selected for high bristle number and five (the lower five) were selected for low bristle numbers. The selection was performed by counting bristles on 100 males and 100 females in each line and then choosing for parents the 20 males and 20 females with the highest and lowest numbers of bristles for their sex. After five gencrations, selection was stopped and the bristles were not counted again until generation 24.

The selection experiment was obviously successful in that the number of bristles increased steadily in the high lines and decreased steadily in the low lines. After five generations, a typical fly in the high lines had about 13 more bristles than the original population, while the low lines had about 7 fewer bristles. This asymmetry in the response to equal selection in both directions is seen in almost all selection experiments.

After five generations, selection was relaxed and the number of bristles moved back toward the number in the original population. This, too, is typical of other selection experiments. One possible explanation is that there is an optimal number of bristles and that natural selection moves the population back toward the optimum once artificial selection is stopped. Another possibility is that the alleles whose frequencies increased in the population as a result of selective breeding have pleiotropic deleterious effects or are linked to deleterious alleles. Once artificial selection stops, natural selection acts on these pleiotropic, or linked, factors.

Our first task is to find a quantitative description of the expected progress of selection in a typical selective breeding experiment. The framework for the discussion is illustrated in Figure 5.5. The starting point, the top curve, is a population of potential parents whose phenotypes are, by assumption, normally distributed. From this population, a group of individuals are chosen to serve as parents for the next generation. The usual practice is to pick a fixed number of individuals to measure and then choose a fixed proportion of these as parents. For example, in the *Drosophila* experiment 100 individuals of each sex were measured and the top and bottom 20 percent were chosen as parents. The chosen parents are then put into male-female pairs and the midparent value of each pair is noted. The mean midparent value of the selected parents, expressed (as always) as the deviation from the population mean, is called the selection differential and is notated by S.

The distribution of the offspring of the selected parents is illustrated in the next line of the figure. As the mean of the selected parents is larger than that of the potential parents, the distribution of the offspring is shifted to the right. The deviation of the offspring mean from the potential parent mean is called the selection response and is notated by R.

What is the relationship between the selection differential, S, and the selection response, R? On page 113 we argued that the regression of offspring on midparent is the heritability. It follows that the mean value of the offspring of a mating with midparent value $P_{MP} = x$ is $h^2 x$. In the selection experiment, there are several pairs of selected parents, each with its own midparent value, say x_i for the ith pair of parents. The expected phenotype of the ith pair of parents is thus $h^2 x_i$. In a selection experiment with n pairs of parents, the

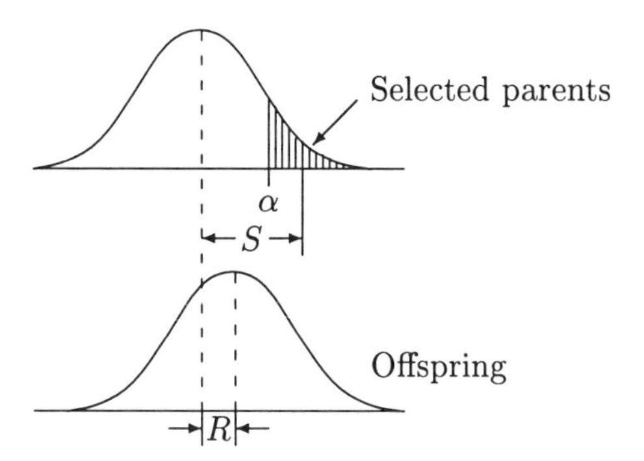

Figure 5.5: The response to selection.

expected phenotype of the offspring is

$$\frac{1}{n}\sum_i E\{P_O \mid P_{MP} = x_i\} = h^2 \frac{1}{n}\sum_i x_i.$$

The term on the left side is the average expected value of the offspring (expressed as a deviation from the parental mean), which is the selection response, R. The sum on the right side is the average value of the phenotypes of the selected parents, or the selection differential, S. In other words, the selection response is simply

$$\boxed{R = h^2 S} \tag{5.11}$$

If the heritability of a trait is one-half, then the response of one generation of selection is to move the population mean halfway between its value in the parental generation and the mean value of the selected parents.

Equation 5.11 is the answer to the question posed at the beginning of this section: How well does selective breeding work? Like the equation for the correlation between relatives, it is a remarkably simple answer to an ostensibly difficult question. However, while $R = h^2 S$ is an accurate description of the response to selection in a single generation, it is not necessarily an accurate predictor of the progress of selection over several successive generations because each generation of selection changes h^2 in ways that are impossible to predict. We will return to this problem after discovering another way to write the selection differential.

When designing a selection experiment, it is often useful to know the proportion of parents needed to obtain a particular selection differential. In fact, the *Drosophila* bristle-number experiment described at the beginning of this section was couched entirely in terms of the proportion selected. We cannot relate that experiment to Equation 5.11 without first finding a connection between the proportion selection and the selection differential. In Figure 5.5, the proportion selected is the area of the shaded portion of the distribution of parental phenotypes. The mean value of the shaded portion is the selection differential. The

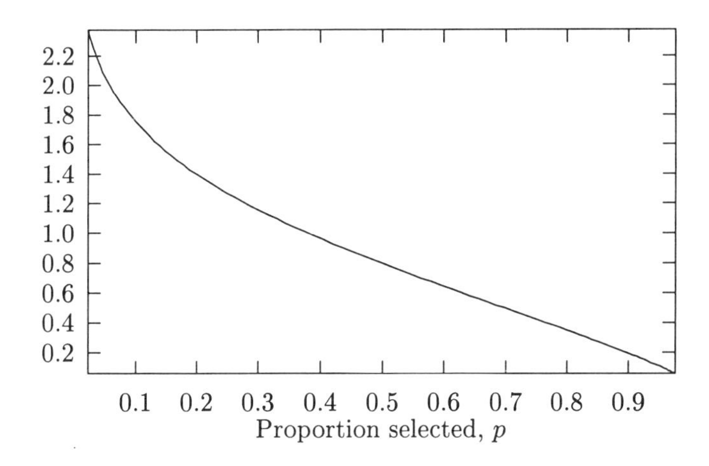

Figure 5.6: The intensity of selection, i, as a function of the proportion of individuals used as parents, p.

connection between the two is given by the formula

$$S = \sqrt{V_P}\, i(p), \tag{5.12}$$

where the function $i(p)$, which is illustrated in Figure 5.6, is called the intensity of selection. The derivation of $i(p)$ is given in Section 5.5. Equations 5.11 and 5.12 together give a new equation for the response to selection,

$$R = h^2 i(p) \sqrt{V_P}, \tag{5.13}$$

which is particularly convenient for many applications.

The 1957 paper by George Clayton, J. A. Morris, and Alan Robertson, which provided Figure 5.4, also provided one of the first demonstrations that Equation 5.13 can accurately describe the response to selection in favorable experimental settings. Clayton et al. did this by estimating the heritability and phenotypic variance of bristle number and then checking how well Equation 5.13 predicts the progress of selection for different fractions of selected individuals. The results are given in Table 5.3 and Figure 5.7. The heritability for bristles is about $h^2 = 0.52$, and the phenotypic variance, an average of the male and female variances, is $V_P = 11.223$, so $\sqrt{V_P} = 3.35$. For the first line in the table, where the intensity of selection is 1.40, the predicted response to selection is

$$R = 0.52 \times 1.40 \times 3.35 = 2.42,$$

which is close to the observed change upward of 2.62 bristles per generation. The response to selection for other values of the intensity of selection in the high lines is also in rough agreement with the theoretical predictions. But what about the low lines? Note that there is nothing in our theory that treats up and down selection differently. Thus, the down line should decrease by about

p	$i(p)$	Predicted	Observed up	Observed down
20/100	1.40	2.42	2.62	1.48
20/75	1.24	2.14	2.20	1.26
20/50	0.97	1.68	1.46	0.79
20/25	0.35	0.61	0.28	−0.08

Table 5.3: The predicted and average observed change in the number of bristles in *Drosophila* in a single generation for different intensities of selection.

2.42 bristles per generation for the highest intensity of selection. In fact, it decreased only by 1.48 bristles. The authors of the paper are at a loss to explain this discrepancy. They do point out that the observed change per generation is obtained by averaging the changes for each of the five generations. In the case of the low lines, they argue that the response to selection declined after the first couple of generations of selection and, by implication, that h^2 declined. However, there is little support for this explanation in Figure 5.4, where the low lines appear to decrease linearly for all five generations of selection. Other possible explanations include the possibility that natural selection opposes artificial selection for lower bristle numbers or that there is a scaling effect due to an asymmetry in the distribution of additive effects on bristle number.

The suggestion that heritabilities decrease during selection points out one of the major obstacles to using Equation 5.13 to predict the response to selection for more than a few generations into the future. Each generation of selection changes the genetic structure of the population by changing allele frequencies and the associations of alleles on chromosomes. As a consequence, the additive variance and the heritability are likely to change and, with them, the response to selection. Unless there is some reason to believe that heritabilities remain relatively constant, perhaps because selection is very weak and mutation is restoring any lost variation, one should not use Equation 5.13 to predict events after the first few generations of selection.

The brevity of this section shows how simply the response to selection follows from the regression of offspring on midparent. However, it belies the considerable complications that arise in actual selection experiments. For a thorough treatment of the response to selection, there is no better place to turn than Douglas Falconer's *Introduction to Quantitative Genetics* (1989).

5.3 Evolutionary quantitative genetics

Quantitative traits pose the same sorts of evolutionary questions as do discrete traits. Just as we puzzled over ubiquitous molecular variation—our Great Obsession—we should wonder why there is so much genetically determined variation in quantitative traits. Similarly, arguments over the strength of selection acting on molecular variation have their counterparts in arguments over the strength of selection acting on quantitative traits. In this section, we will just brush the surface of these questions, which are not nearly as well posed for quantitative traits as they are for molecular variation because we do not un-

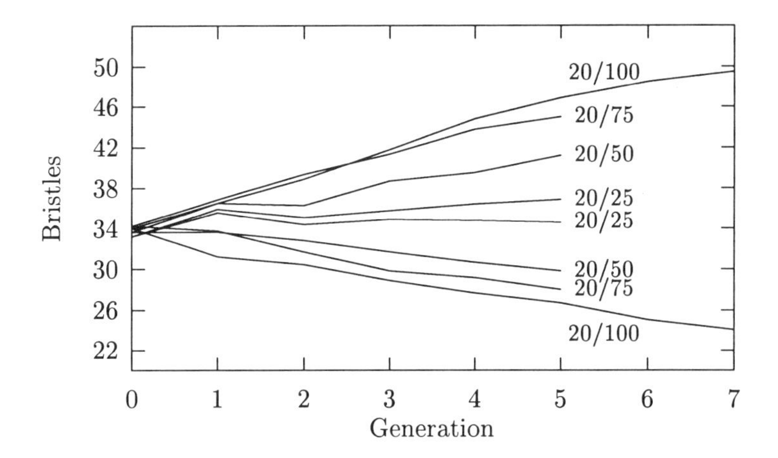

Figure 5.7: The results of selection of different intensities for bristle number in *Drosophila.* The upper four curves are averages of three to five lines selected for higher numbers of bristles with the proportion selected as given in the figure. The lower four curves are for selection for fewer numbers of bristles. The data are from Clayton et al. (1957).

derstand the genetic basis of quantitative variation. This difference is offset somewhat by the fact that quantitative traits are under much stronger selection than are molecular traits, making it easier to guess the adaptive value of many quantitative traits.

Quantitative genetics has been used to reconstruct the selective forces acting on quantitative traits. The basic idea is to turn our previous discussion on its head by using the selection response to estimate the selection differential. An interesting example is in a paper by Russ Lande (1976), which examines selection on paracone height in fossil horses. In the course of their radiation from the Eocene to the Pleistocene, horses moved from eating leaves in forests to eating grasses in plains. As eating grasses wears down teeth faster than eating leaves, horses evolved higher ridges, called paracones, on their teeth. Table 5.4 summarizes some of the relevant facts of the evolution of paracone height. The change in paracone height over an interval of time of length T is called R_T. The ratio of R_T to the phenotypic standard deviation, $\sqrt{V_P}$, for pairs of species is given in the second column of Table 5.4. The time of separation of the pairs is given in the third column.

Let us assume that the increase in paracone height was due to a constant directional selection pressure and try to find the strength of selection. The selection response may be calculated by first rearranging Equation 5.13 to

$$\frac{R}{\sqrt{V_P}} = h^2 i(p).$$

The response per generation, assuming that a horse generation is one year, is

From species to species	$R_T/\sqrt{V_P}$	T	$1-p$
Hyracotherium to *Mesohippus*	10.6	10×10^6	4×10^{-7}
Mesohippus to *Merychippus*	25.6	5×10^6	2×10^{-6}
Merychippus to *Neohippus*	7.8	1.75×10^6	2×10^{-6}
Hyracotherium to *Neohippus*	44.0	16.75×10^6	1×10^{-6}

Table 5.4: The fraction of selective deaths, p, required to account for the evolution of paracone height in horses under the assumption of directional selection.

obtained by dividing the total difference in paracone height by the length of separation, T, $R = R_T/T$. The heritability of paracone height is assumed to be one-half, a typical figure for morphologic traits. Thus, the intensity of selection becomes

$$i(p) = \frac{2R_T}{T\sqrt{V_P}}.$$

The parameters on the right side may be obtained from Table 5.4 to arrive at the intensity of selection. For example, from the first line the intensity of selection is

$$i(p) = \frac{2 \times 10.6}{10^7} = 2.12 \times 10^{-6}.$$

The value of p corresponding to the intensity of selection could, in principle, be obtained from Figure 5.6. However, the resolution of the graph for such a small intensity of selection is not sufficient. The values of $1 - p$, obtained by Lande using approximations to the integrals appearing in the definition of the intensity of selection, are in the final column of Table 5.4.

The values of $1-p$, which are the proportion of horses not included as parents each generation, suggest that only about one to four horses of every million need suffer a selective death in order to evolve paracone height at the rate seen in the fossil record. This is incredibly weak selection! So weak, in fact, that genetic drift could well dominate selection. Does this mean that we should entertain drift as a candidate to explain the evolution of paracone height? Perhaps, but a much more profitable direction is to accept that our assumption of constant directional selection is unrealistic. Paracone height is probably under stabilizing selection for an optimum height that is determined by, among other factors, the time spent feeding on grasses and the nature of the grasses. The optimum itself will slowly change upward as the grasslands begin to dominant the landscape. Instead of a picture of an excruciatingly slow response to a distant selective goal, we picture very strong selection keeping the species right at the optimum paracone height. Under this hypothesis, the slow increase in paracone height is due to the slow change in the optimum rather than to a slow response to very weak selection.

The investigation of paracone height poses the questions: How fast have quantitative traits evolved in the fossil record? How do these rates of evolution compare to rates observed under artificial selection? Is the rate of evolution of

	Number	Time interval	Rate (Darwins)
Selection experiments	8	3.7 yr	58700
Recent colonizations	104	170 yr	370
Post-Pleistocene mammals	46	8200 yr	3.7
Fossil vertebrates	228	1.6 MY	0.08
Fossil invertebrates	135	7.9 MY	0.07

Table 5.5: Rates of morphological evolution. The time intervals are measured in either years (yr) or millions of years (MY). The table is from Maynard Smith (1989).

paracone height typical or unusually fast or slow? J. B. S. Haldane, one of the pioneers of population genetics, proposed that rates of phenotypic evolution be quantified by

$$k_h = \log(x_2/x_1)/T \text{ Darwins.}$$

In this expression, x_1 and x_2 are measures of some trait for two species on the same lineage that are separated by T million years. The units of k_h are called Darwins. For example, the paracone height of *Hyracotherium* is $x_1 = 1.54$ and of *Mesohippus* is $x_2 = 2.12$. As these two horses are separated by 20 million years, the rate of paracone evolution is

$$k_h = \frac{\log(2.12/1.54)}{20} = 0.016 \text{ Darwins.}$$

Table 5.5 lists some summary rates of evolution for various situations.

The most striking feature of the table is that the rate of evolution in selection experiments is at least two orders of magnitude faster than any rate seen in nature or the fossil record. The fastest rates in nature are for recent colonizers, which might be expected to evolve quickly to adapt to their new environments. Paracone height is clearly evolving very slowly compared to what is possible in the laboratory or even what is seen as the fastest rates in nature. This lends support to the view that the evolution of paracone height is due to a slowly changing optimum rather than directional evolution to a radical new environment. Thus, when considering the applicability of selection theory to the fossil record, one should never assume that the selective forces have remained constant for appreciable periods of time.

The Lande paper is a quantitative genetics analogue to the study of rates of molecular evolution. Similarly, there is a quantitative genetics version of the Great Obsession: Why is there so much genetic variation for quantitative traits? The study of quantitative genetic variation is very similar to that of molecular variation. As a first step, for example, one might try to find the amount of quantitative variation introduced each generation by mutation and the amount removed by genetic drift to see if the equilibrium determined by these two forces is compatible with observed levels of quantitative variation, just as Kimura and Ohta did for molecular variation.

A seminal paper on quantitative genetic variation is a short work by George Clayton and Alan Robertson published in 1955. Their main interest was the extent to which mutations may contribute to the overall response to selection in experiments like that of Clayton, Morris, and Robertson described in the previous section. Before their paper, the relative roles of allele frequency change and mutational input to the selection response were not well understood theoretically or empirically. We begin by first describing the experiment and its analysis, and then we will discuss its implications for the question of the maintenance of variation.

The experiment used a straightforward selective breeding protocol. The only novel feature was the initial population, which was nearly homozygous because of many generations of brother-sister mating. With so little initial variation, the response to selection should be due, in large part, to newly arising mutations affecting bristle number. High and low lines were selected by choosing the 10 most extreme individuals out of 25 for each sex, which gives $\imath(0.4) = 0.94$ for the intensity of selection. Such selection continued for 14 generations. By averaging the high and low lines across all 14 generations, Clayton and Robertson concluded that the average response in one generation of selection was $R = 0.027$ bristles. Using Equation 5.13 and their estimate that $V_P = 4.41$, the average additive variance over the 14 generations is

$$\overline{V}_A = 0.06,$$

where the bar over V_A is a reminder that this is an average over generations.

In the course of the experiment, genetic drift removes additive variance at a rate of $1/(2N)$, and mutation augments the additive variance by an amount V_m each generation. V_m is the quantity that we want to estimate. Thus, the model for the evolution of V_A during the 14 generations is,

$$V_A(t) = \left(1 - \frac{1}{2N}\right) V_A(t-1) + V_m, \tag{5.14}$$

which appears in the text of the paper at the top of page 155, albeit with different notation. We will not derive this equation, although it should be in accord with your intuition. In each generation, genetic drift reduces the additive variance by the factor $1-1/(2N)$, just as heterozygosity is reduced by this factor when discussing molecular variation. In each generation, mutation augments the additive variance by V_m, which is analogous to the mutational input, $2Nu$, of molecular variation.

Equation 5.14 can be solved iteratively, once we know the additive variance that was present at the start of the experiment, $V_A(0)$. The authors chose not to estimate $V_A(0)$ directly, but rather to argue that additive variance is at an equilibrium between its removal by inbreeding (brother-sister mating) and its introduction by mutation. Brother-sister mating removes variance at the rate 0.191. (The calculation of this number may be found on page 260 of Dan Hartl and Andy Clark's 1989 textbook.) Thus, at equilibrium

$$0.191 V_A(0) = V_m,$$

or $V_A(0) = 5.2V_m$.

Equation 5.14 may now be iterated to find the additive variance in successive generations. For generation 1,

$$V_A(1) = 0.975 \times 5.2V_m + V_m = 6.07V_m,$$

using a population size of $N = 20$, for which

$$1 - \frac{1}{2N} = 0.975.$$

The next generation is

$$V_A(2) = 0.975 \times 6.07V_m + V_m = 6.92V_m.$$

Continuing in this way, $V_A(14) = 15V_m$. The average additive variance (add up the $V_A(i)$ and divide by 14) is

$$\overline{V}_A = 10V_m = 0.06,$$

from which we conclude that $V_m = 0.006$. In a typical *Drosophila* population, $V_A = 5$. Thus, mutation augments the additive variance by about 0.006/5, or 0.1 percent its value each generation.

Clearly, the mutational input to this particular quantitative trait each generation is very small. To account for the observed value of V_A in natural populations, we must invoke a very weak force to remove the variation. Genetic drift is always a good candidate for a weak force, providing that we are willing to accept that bristle number is a neutral trait. If so, we can use Equation 5.14 to argue that the equilibrium additive variance will satisfy

$$\hat{V}_A = \left(1 - \frac{1}{2N}\right)\hat{V}_A + V_m,$$

which gives

$$\hat{V}_A = 2NV_m.$$

Recalling that $V_m/V_A = 0.001$, we must conclude that the effective size of *Drosophila* is 500, an absurdity. The additive variance in natural populations is far less than that predicted by this model.

Where did we go wrong? Many evolutionists would argue that bristle number is not a neutral trait. Rather, there is an optimum number of bristles, and mutations that move a fly away from the optimum are selected against. This would have the effect of lowering the additive variance well below that predicted by the neutral models. Others would argue differently. They would say that bristle number really is a neutral trait, but that mutations changing the numbers of bristles affect other traits as well, and these other traits are under selection. In this case, there is selection against bristle mutations because of their pleiotropic effects. Currently, we do not know which of these two explanations is closer to reality; perhaps neither is accurate, and some other set of forces is maintaining additive variance.

5.4 Dominance

This final section introduces an approach to quantitative genetics based on only two alleles at a locus rather than the large number of alleles used in Section 5.1. There are two immediate benefits. The first is that dominance may be included in a way that more nearly fits our biological sense of dominance than would be possible with the statistical approach of the first section. The second is that we will be able to show that the genetic contribution to traits affected by many di-allelic loci, each of small effect, is approximately a normal distribution. Thus, this section could be viewed as a genetically based justification for the approach taken in the first section.

With dominance, the linear model for the construction of a phenotype is

$$P = X_m + X_p + X_{mp} + \mathcal{E},$$

where the new term, X_{mp}, captures the dominance relationships between the maternally and paternally derived alleles. If V_D is the variance in X_{mp}, then, assuming that the additive and dominance contributions are uncorrelated,

$$V_P = V_A + V_D + V_E.$$

There is no obvious reason why the additive and dominance contributions should not be correlated. In fact, a little reflection shows that we haven't said precisely what is meant by the additive and dominance contributions. The confusion will be cleared up in this section. We will show that the additive effects are chosen in such a way as to maximize their contributions to the phenotype while assuring that the additive effects are uncorrelated with the dominance effects.

To start, we need to define the contributions to the phenotype made by a single locus with two alleles:

Genotype:	A_1A_1	A_1A_2	A_2A_2
Frequency:	p^2	$2pq$	q^2
Genotypic value:	a_{11}	a_{12}	a_{22}
Additive value:	$2\alpha_1$	$\alpha_1 + \alpha_2$	$2\alpha_2$

The genotypic values are the contributions of the genotypes to the phenotype expressed as deviations from the population mean. Consequently, the mean of the genotypic values must be zero,

$$p^2 a_{11} + 2pq a_{12} + q^2 a_{22} = 0.$$

The additive values of the three genotypes are values that come as close as possible to making this locus one with no dominance. The additive values depend of the genotypic values and the allele frequencies. They will be derived shortly. For now, we will note only that the additive values are expressed as deviations from the population mean, so the mean of the additive effects must be zero as well,

$$p^2 2\alpha_1 + 2pq(\alpha_1 + \alpha_2) + q^2 2\alpha_2 = 2\alpha_1 p + 2\alpha_2 q = 0. \tag{5.15}$$

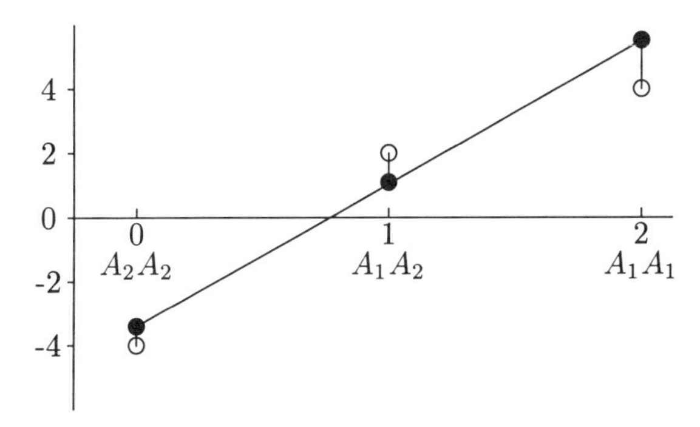

Figure 5.8: An illustration of the relationship between genotypic values, the open circles, and additive values, the closed circles, for the case $a_{11} = 4$, $a_{12} = 2$, $a_{22} = -4$, and $p = 0.38197$.

The variance of the genetic contribution to the phenotype, the genetic variance, is

$$V_G = p^2 a_{11}^2 + 2pq a_{12}^2 + q^2 a_{22}^2, \tag{5.16}$$

recalling that the mean the genotypic values is zero. The additive variance is given by

$$\begin{aligned} V_A &= p^2(2\alpha_1)^2 + 2pq(\alpha_1 + \alpha_2)^2 + q^2(2\alpha_2)^2 \\ &= 2p^2\alpha_1^2 + 2(p\alpha_1 + q\alpha_2)^2 + 2q\alpha_2^2 \\ &= 2(p\alpha_1^2 + q\alpha_2^2), \end{aligned} \tag{5.17}$$

where the derivation of the final form requires Equation 5.15. At this point, however, we do not know the additive values (α_i is function of a_{ij} and p); we will return to V_A once we do.

The method for finding the additive values is illustrated in Figure 5.8. The horizontal axis is the number of A_1 alleles in the genotype. With each new A_1 allele, the additive value of the genotype is increased by α_1, as illustrated by the sloping line. The deviation of the genotypic values from the additive values is illustrated by the vertical lines connecting the two. The equation for the line is obtained by minimizing the mean squared deviation of the line from genotypic values,

$$Q(\alpha_1, \alpha_2) = p^2(a_{11} - 2\alpha_1)^2 + 2pq(a_{12} - \alpha_1 - \alpha_2)^2 + q^2(a_{22} - 2\alpha_2)^2,$$

coupled with the previous assumption that the mean of the additive effects is zero. Finding a best-fit line by minimizing the squared deviation is called the method of least squares. The advantage of this particular method is that the additive and dominance effects are uncorrelated, as you will show in a problem once you have all of the requisite formulae.

The additive effects are found by minimizing $Q(\alpha_1, \alpha_2)$ using the usual method of setting the derivatives of Q equal to zero,

$$\frac{\partial Q}{\partial \alpha_1} = -4[p^2(a_{11} - 2\alpha_1) + pq(a_{12} - \alpha_1 - \alpha_2)] = 0$$
$$\frac{\partial Q}{\partial \alpha_2} = -4[pq(a_{12} - \alpha_1 - \alpha_2) + q^2(a_{22} - 2\alpha_2)] = 0.$$

Both equations are readily simplified by first isolating the mean additive effect, $p\alpha_1 + q\alpha_2$, and setting it equal to zero and then dividing the first equation by $-4p$ and the second equation by $-4q$. The resulting equations may be solved to obtain

$$\alpha_1 = pa_{11} + qa_{12} \tag{5.18}$$
$$\alpha_2 = pa_{12} + qa_{22}. \tag{5.19}$$

Problem 5.5 *Confirm that the mean additive effect is zero using the above expressions for* α_1 *and* α_2.

The additive values may now be substituted into Equation 5.17 to obtain the additive variance in terms of the genotypic values,

$$V_A = 2pq[p(a_{11} - a_{12}) + q(a_{12} - a_{22})]^2. \tag{5.20}$$

Finally, the dominance variance is obtained by subtracting the additive variance from thc genetic variance,

$$V_D = V_G - V_A = p^2q^2(a_{11} - 2a_{12} + a_{22})^2. \tag{5.21}$$

Thus, we have completed our partitioning of the total genetic variance into additive plus dominance components,

$$V_G = V_A + V_D. \tag{5.22}$$

Problem 5.6 *Show that a locus without dominance has* $V_D = 0$. *Construct an example of a locus with* $V_A = 0$ *but* $V_D > 0$.

The next step is to find the genetic contribution to the covariance between an arbitrary pair of relatives. The approach is mundane, but the algebra is cumbersome. Recall from page 161 that the covariance between two random variables is

$$\sum_i \sum_j p_{ij} x_i y_j.$$

In our case, the random variables are the genotypic values of relatives X and Y as summarized in Table 5.6. For example, the first line gives the pertinent data for a pair of relatives, both of whom are A_1A_1. The contribution of this locus to the phenotype of both relatives is a_{11}. The frequency of this pair is the probability that X is A_1A_1, p^2, times the probability that Y is A_1A_1 given

Genotypes		Values		Frequency
X	Y	X	Y	
A_1A_1	A_1A_1	a_{11}	a_{11}	$p^2[r_0p^2 + r_1p + r_2]$
A_1A_1	A_1A_2	a_{11}	a_{12}	$p^2[r_02pq + r_1q]$
A_1A_1	A_2A_2	a_{11}	a_{22}	$p^2[r_0q^2]$
A_1A_2	A_1A_1	a_{12}	a_{11}	$2pq[r_0p^2 + r_1p/2]$
A_1A_2	A_1A_2	a_{12}	a_{12}	$2pq[r_02pq + r_1/2 + r_2]$
A_1A_2	A_2A_2	a_{12}	a_{22}	$2pq[r_0q^2 + r_1q/2$
A_2A_2	A_1A_1	a_{22}	a_{11}	$q^2[r_0p^2]$
A_2A_2	A_1A_2	a_{22}	a_{12}	$q^2[r_02pq + r_1p]$
A_2A_2	A_2A_2	a_{22}	a_{22}	$q^2[r_0q^2 + r_1q + r_2]$

Table 5.6: All possible pairs of relatives needed for the calculation of the covariance between relatives.

that X is A_1A_1. The latter probability is sum of the probabilities of three mutually exclusive events. The first event is when X and Y share no identical-by-descent alleles, r_0, times the probability that Y is A_1A_1 given that it shares no identical-by-descent alleles with X, p^2. The probabilities of the other two events are obtained in a similar fashion, as are the rest of the frequencies in the table.

The covariance is calculated by adding up nine terms, each of which is the frequency of a pair times the product of the two values of that pair. The sum is simplified by considering, in turn, the coefficients of the r_i. The coefficient of r_0 is

$$(p^2a_{11} + 2pqa_{12} + q^2a_{22})^2,$$

which is just the square of the average genotypic value, which is zero.

The coefficient r_1, after a modest bit of algebra that requires the use of equation 5.18, is

$$p\alpha_1^2 + q\alpha_2^2,$$

which, by Equation 5.17, is $V_A/2$. Finally, the coefficient of r_2 is, from Equation 5.16, V_G. Taken together, these calculations show that

$$\mathrm{Cov}\{P_X, P_Y\} = (r_1/2)V_A + r_2V_G,$$

or

$$\begin{aligned}\mathrm{Cov}\{P_X, P_Y\} &= (r_1/2 + r_2)V_A + r_2V_D \\ &= rV_A + r_2V_D. \end{aligned} \tag{5.23}$$

When there is no dominance, this is the same as Equation 5.8, which was obtained by a different method. In fact, the two derivations share one critical feature: they both make explicit use of the resemblance that comes from sharing identical-by-descent alleles.

Genotype	Frequency	X_m	X_p	X_{mp}
A_1A_1	p^2	α_1	α_1	$a_{11} - 2\alpha_1$
A_1A_2	pq	α_1	α_2	$a_{12} - \alpha_1 - \alpha_2$
A_2A_1	pq	α_2	α_1	$a_{12} - \alpha_2 - \alpha_1$
A_2A_2	q^2	α_2	α_2	$a_{22} - 2\alpha_2$

Table 5.7: All possible ways to make a genotype.

Problem 5.7 *Suppose the correlation between full sibs is 5/24 and that between half sibs is 1/12. If the phenotypic variance is 6, what are the values of V_A and V_D?*

In Section 5.1, we assumed that there was no dominance and used the model

$$P = X_m + X_p + \mathcal{E}$$

to partition variances and find the correlation between relatives. With dominance, the model was extended to

$$P = X_m + X_p + X_{mp} + \mathcal{E}, \tag{5.24}$$

where the new term, X_{mp}, represents the contribution of dominance to the phenotype. As with X_m and X_p, X_{mp} is a population quantity that depends on allele frequencies. With just one locus and two alleles, it is a simple matter to describe the universe of genetic effects, as is done in Table 5.7. Furthermore, we can partition the phenotypic variance into

$$V_P = V_A + V_D + V_E,$$

providing that we assume no genotype-environment interactions. It appears that we also assumed that the additive and dominance effects are uncorrelated because there is no covariance term for these factors. However, the least-squares method used to find the additive effects makes the additive and dominance contributions uncorrelated.

Problem 5.8 *Prove that the covariance between the additive and dominance effects is zero.*

Up to this point, both of our models assume that the genetic contribution to a trait is due to a single locus. Obviously, most quantitative traits will be affected by more than one locus. The generalization to many loci for di-allelic loci comes from simply rewriting Equation 5.24 as

$$P = \sum_i [X_m(i) + X_p(i) + X_{mp}(i)] + \mathcal{E}, \tag{5.25}$$

where $X_m(i)$, $X_p(i)$, and $X_{mp}(i)$ are the genetic contributions from the ith locus and the sum is over all loci that affect the trait. The variance in the phenotype is

$$V_P = V_A + V_D + V_E,$$

where now the additive variance is defined as the sum of the additive variances of the individual loci,

$$V_A = \sum_i V_A(i).$$

The dominance variance is summed across loci as well. Once again, we have slipped in an assumption to remove covariance terms. This time, we assume that there is no interaction between alleles at different loci. If there were an interaction, the additional variance due to the interaction would be called the epistatic variance, V_I, and the phenotypic variance would be

$$V_P = V_A + V_D + V_I + V_E.$$

In this more general setting, the heritability is still defined as

$$h^2 = \frac{V_A}{V_P}$$

and is often called the narrow sense heritability. The broad sense heritability is the ratio of all of the genetic variances to the phenotypic variance,

$$H^2 = \frac{V_A + V_D + V_I}{V_P}.$$

The covariance between relatives with many noninteracting loci is

$$\mathrm{Cov}\{P_X, P_Y\} = rV_A + r_2 V_D,$$

where, as with the variance partitions, the additive and dominance variances are the sums of the variances of the individual loci. Thus, as far as variances and covariances go, the one-locus and multilocus models give the same results.

There is one very important property of the two-allele model that does change when more loci are added: The distribution of the genetic effects approaches a normal distribution. Notice that the distribution of height in the left side of Figure 5.1 approximates that of a normal distribution, albeit crudely because of the small sample size. There is no a priori reason why the phenotypic distribution should be so, well, normal. The Central Limit Theorem from probability theory does provide a partial explanation. This theorem states that the distribution of the sum of independent random variables, suitably scaled, approaches a normal distribution as the number of elements in the sum increases. The approach to the normal distribution is usually quite fast. For example, the distribution of the sum of as few as 10 uniform random variables looks remarkably like a normal distribution. Thus, if as few as 10 loci contributed to a trait, the genetic contribution would look normal.

Of course, the phenotypic distribution also has an environmental component that must itself be approximately normally distributed if the phenotypic distribution is to be normally distributed. In fact, this appears to be generally true as judged from an examination of the phenotypic distribution of individuals that are genetically identical, as occurs, for example, in inbred lines. Perhaps the environmental component is also the sum of many small random effects that add to produce their effects on the phenotype.

One of the outstanding problems of quantitative genetics concerns the number of loci that contribute to the variation in a trait. Speculation has ranged from a very large number of loci (tens to hundreds), each with a very small and roughly equivalent effect, to very few loci (one to five), each with a relatively large effect. Direct genetic analysis of some traits (e.g., the number of bristles on a Drosophila), supports the latter view. With modern molecular techniques, this issue should be resolved in the near future.

5.5 The intensity of selection

This short section provides a quick derivation of the intensity of selection. The derivation uses properties of the normal distribution, which are given on page 164.

The probability density of parental phenotypes is the normal density with mean zero and variance V_P,

$$\frac{1}{\sqrt{2\pi V_P}}\exp(\frac{-x^2}{2V_P}).$$

Thus, the proportion that is selected is the area

$$p = \int_\alpha^\infty \frac{1}{\sqrt{2\pi V_P}}\exp(\frac{-x^2}{2V_P})dx,$$

where α is the largest number that is less than the values of the selected parents. The proportion may be simplified by changing the variable of integration to $y = x/\sqrt{V_P}$,

$$p = \int_{\alpha/\sqrt{V_P}}^\infty \frac{1}{\sqrt{2\pi}}e^{-y^2/2}dy. \tag{5.26}$$

The selection differential is the mean of the selected parents, which is the mean of the parents found in the shaded portion of Figure 5.5. The density of the selected parents is the truncated normal density

$$\frac{C}{\sqrt{2\pi V_P}}\exp(\frac{-x^2}{2V_P}),\ x > \alpha,$$

where the constant C is chosen to make the integral of the density equal to one,

$$C = \left(\int_\alpha^\infty \frac{1}{\sqrt{2\pi V_P}}\exp(\frac{-x^2}{2V_P})dx\right)^{-1}.$$

The mean of the selected parents is thus

$$S = C \int_{\alpha}^{\infty} \frac{x}{\sqrt{2\pi V_P}} \exp(\frac{-x^2}{2V_P}) dx.$$

As with the proportion selected, the selection differential may be simplified by changing the variable of integration to $y = x/\sqrt{V_P}$,

$$S = \sqrt{V_P}\, i(\alpha/\sqrt{V_P}), \tag{5.27}$$

where

$$i(\alpha/\sqrt{V_P}) = \frac{\int_{\alpha/\sqrt{V_P}}^{\infty} y e^{-y^2/2} dy}{\int_{\alpha/\sqrt{V_P}}^{\infty} e^{-y^2/2} dy} \tag{5.28}$$

is called the intensity of selection.

Notice that the proportion selected, as given by Equation 5.26, and the intensity of selection, as given by Equation 5.28, are both functions of the quantity $\alpha/\sqrt{V_P}$. In fact, they are both monotonic functions of $\alpha/\sqrt{V_P}$; p is a decreasing function and S is an increasing function. As a consequence, for each value of p there is a unique corresponding value of the intensity of selection. Thus, we can view the intensity of selection as a function of p rather than as a function of $\alpha/\sqrt{V_P}$. Unfortunately, the functional relationship cannot be written as a simple formula because both p and $i(\alpha/\sqrt{V_P})$ are functions of integrals. However, it is easy to evaluate the integrals with a computer and in so doing obtain Figure 5.6. This figure relates the intensity of selection to the proportion of selected individuals. The selection differential is then obtained using Equation 5.27, which should now be written as

$$S = \sqrt{V_P}\, i(p),$$

which is the same as Equation 5.12.

The approach used here to obtain the selection differential uses an implicit assumption that there are a large number of measured individuals that are used to select the parents. If the number of measured individuals is 20 or fewer, then a more accurate intensity-of-selection function is required. Such functions may be found in any standard text on quantitative genetics.

5.6 Answers to problems

5.1 The heritability is $h^2 = 2/(2+3) = 0.4$. The covariance of parent and offspring is $0.4/2 = 0.2$.

5.2 If $V_A = 2$ and $V_E = 3$, then $h^2 = 2/3$. The coefficient of relationship between parent and offspring is 1/2, so their correlation is $1/2 \times 2/3 = 1/3$. The correlations between the other relatives are found in the same manner.

5.3 The heritability of height is $\sqrt{2} \times 0.476 = 0.673$. Thus, about 67 percent of all of the variation in height in the class is attributable to genetic causes, and the other one-third is environmental in origin. The midparent height is $x = (3 + 1)/2 = 2$; the expected height of their child is

$$h^2 x = 0.673 \times 2 = 1.35 \text{ inches.}$$

5.4 Using a pocket calculator, the correlation between parent and offspring is 0.4366. From Table 5.2 we have $h^2 = 2 \times 0.4336 = 0.8672$.

5.5 The mean additive effect is

$$p\alpha_1 + q\alpha_2 = p^2 a_{11} + 2pq a_{12} + q^2 a_{22} = 0$$

by the assumption that the mean of the a's is zero.

5.6 No dominance means that $a_{12} = (a_{11} + a_{22})/2$, which, when plugged into Equation 5.21, gives $V_D = 0$. For the second part, $a_{11} = a_{22} = -a_{12} = a < 0$ and $p = 1/2$ gives $V_A = 0$ and $V_D > 0$. Note that, if the a were fitnesses, this case would correspond to an overdominant locus at equilibrium.

5.7 Equation 5.23 shows that the correlation between relatives is

$$\text{Corr}\{P_X, P_Y\} = rh^2 + r_2 \frac{V_D}{V_P}.$$

As the correlation between half-sibs is 1/12, $h^2 = V_A/V_P = 1/3$. Using this with the correlation between full sibs we get $V_D/V_P = 1/6$. Multiplying both by $V_P = 6$ gives $V_A = 2$ and $V_D = 1$.

5.8 From Table 5.7, we find that the covariance between additive and dominance effects is

$$\begin{aligned}
\text{Cov}\{P_X, P_Y\} &= p^2 2\alpha_1(a_{11} - 2\alpha_1) \\
&\quad + 2pq(\alpha_1 + \alpha_2)(a_{12} - \alpha_1 - \alpha_2) + q^2 2\alpha_2(a_{22} - 2\alpha_2) \\
&= -[p^2(2\alpha_1)^2 + 2pq(\alpha_1 + \alpha_2)^2 + q^2(2\alpha_2)^2] \\
&\quad + 2p\alpha_1(pa_{11} + qa_{12}) + 2q\alpha_2(pa_{12} + qa_{22}) \\
&= -V_A + V_A = 0.
\end{aligned}$$

Chapter 6

The Evolutionary Advantage of Sex

In this chapter, we use our knowledge of population and quantitative genetics to explore one the most compelling of evolutionary conundrums: Why sex? Rather than attacking all aspects of this question, we restrict our attention to sex in eukaryotic anisogamous creatures, species where the female produces relatively large gametes, such as eggs or seeds, and the male produces a plethora of small and mobile gametes, such as sperm or pollen. This describes, in other words, all of the animals and higher plants, most of which indulge in sexual reproduction. However, there are species that do not. Most could be called parthenogenetic. There are many forms of parthenogenesis. In forms where meiosis is not involved, like apomixis and endomitosis, offspring are genetically almost identical to their parent. The differences that do occur are caused by mutation or mitotic recombination. In forms of parthenogenesis where meiosis is involved, like automixis, offspring are not identical to their parent and are usually inbred. Here we will be concerned only with the first form of parthenogenesis, which may be found in many major groups of plants and animals.* However, except for one group of rotifers, there are no major taxonomic groups that are entirely parthenogenetic. Parthenogenetic species must inevitably face quick extinction due, presumably, to problems associated with the absence of sex, despite the fact that parthenogenetic species enjoy a twofold advantage over their sexual siblings.

Figure 6.1 illustrates the twofold cost to sexual reproduction in anisogamous species in stable populations. Each female, whether sexual or not, will leave behind, on average, two offspring. On average, in sexual species, one of the two offspring will be a female and the other a male. Each of these offspring carries one haploid complement of the female's genetic material; the other haploid genome comes from the male parent. A parthenogenetic female will also leave

*Interestingly, parthenogenesis has not been found in natural populations of mammals or birds.

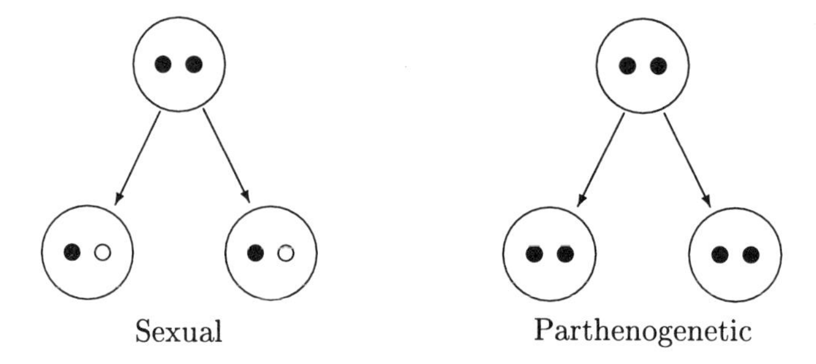

Figure 6.1: A comparison of the consequences of sexual versus parthenogenetic reproduction. The female parent is illustrated in both cases. The filled circles in the offspring come from the mother; the father's genetic contribution is indicated by the open circles in the sexual species.

behind two offspring, but both will be females because they are clones of their mother. Moreover, all of the genetic material, four haploid complements, comes from the mother. Thus, the parthenogenetic female leaves behind twice as much of her genetic material as does the sexual female.

Imagine a sexual species in which a dominant mutation arises that causes parthenogenesis. The genome carrying the mutation will appear in two individuals in the next generation, four the generation after that, and so on, doubling in number each generation. Meanwhile, all genomes without the mutation will be replacing themselves once each generation, on average. Eventually, the species will be made up entirely of parthenogenetic individuals. This is the twofold cost of sexual reproduction.

There must be some problem with this scenario or else all species would become parthenogenetic. One possibility, of course, is that mutations to parthenogenesis either do not occur or are so deleterious that they swamp their twofold advantage. However, in many groups parthenogenesis does recur, so we must look elsewhere for an explanation of its ultimate failure.

Sex in diploids is associated with segregation and meiotic crossing-over; parthenogenetic species have neither. Thus, a natural place to look for the advantages of sex is the examination of segregation and recombination. Both, as we shall see, can confer certain advantages to a species.

6.1 Genetic segregation

Although the reasons for sex have been discussed for decades, most of the arguments centered on crossing-over and recombination. Only recently did Mark Kirkpatrick and Cheryl Jenkins point out that segregation itself can confer an advantage. Their 1989 paper, "Genetic segregation and the maintenance of sexual reproduction," points out that the substitution of an advantageous mutation in a parthenogenetic species requires two mutations in a lineage, whereas substitution in a sexual species requires only one mutation. If the evolutionary

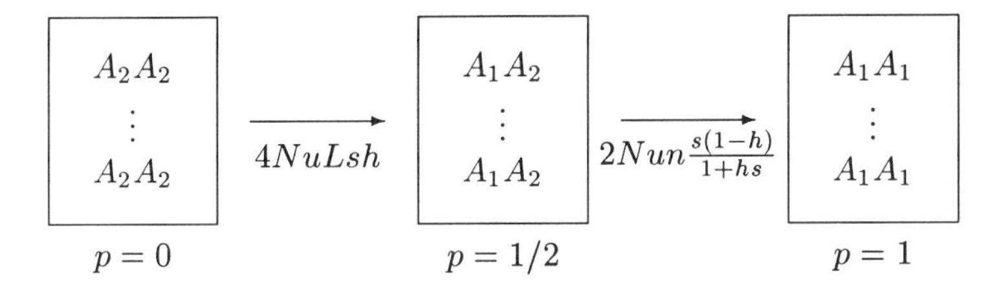

Figure 6.2: Evolution in parthenogens.

success of a species depends on its ability to evolve rapidly and if advantageous mutations are relatively rare, then it is easy to imagine that the parthenogens will be at a disadvantage.

While plausible, this argument carries little force unless it can be shown to enhance the lot of sexual species sufficiently to overcome their twofold cost of sex. Kirkpatrick and Jenkins did this by first finding the number of loci in a parthenogenetic species that are heterozygotes waiting for a second mutation that allows the fixation of the advantageous mutation and then calculating the cost of being heterozygous at these loci.

With incomplete dominance, the fixation of an advantageous mutation in a parthenogen occurs in two steps, as illustrated in Figure 6.2. In the beginning, a typical locus is imagined to be fixed for the A_2 allele with the fitnesses of the genotypes as follows:*

Genotype:	A_1A_1	A_1A_2	A_2A_2
Fitness:	$1+s$	$1+hs$	1

The A_1 allele is favored, so $0 < s$. The rate of mutation to the A_1 allele is u. Each generation, $2Nu$ A_1 mutations enter the population, on average, and a fraction $2sh$ of these escape loss by genetic drift. (In Section 3.7 we showed that the probability of not being lost due to the action of genetic drift is twice the selective advantage of the heterozygote.) Thus, the rate of entry of A_1 alleles into the population is $4Nuhs$. If there are L homozygous loci experiencing this sort of selection, then the rate of conversion from homozygotes to heterozygotes is

$$4NuLhs. \tag{6.1}$$

The environment is assumed to be changing in such a way that L is constant rather than decreasing with the substitution of each advantageous mutation.

The second stage is the conversion of A_1A_2 heterozygotes to A_1A_1 homozygotes, which requires the fixation of an A_1 allele on the chromosomes with the A_2 allele. The mean number of A_1 mutations entering each generation at one

*The notation for fitness will deviate slightly from that of earlier chapters in order to approximate that of Kirkpatrick and Jenkins' paper. Population genetics suffers from a lack of notational conventions; now you will get a taste of what will occur when reading the original literature.

locus is Nu. (The factor of two is missing because each individual is an A_1A_2 heterozygote.) The survival probability of these mutations is twice the selective advantage of A_1A_1 over A_1A_2. The selection coefficient of an A_1A_1 homozygote relative to the A_1A_2 heterozygote, call it s', is found by solving

$$1 + s' = \frac{1+s}{1+hs}$$

to obtain $s' = s(1-h)/(1+hs)$. Thus, the rate of conversion of a particular locus from a population of heterozygotes to one of homozygotes is

$$2Nus(1-h)/(1+hs).$$

If n loci are currently heterozygous, then the rate of conversion from A_1A_2 to A_1A_1 for the genome is

$$2Nun\frac{s(1-h)}{1+hs}. \tag{6.2}$$

The equilibrium number of heterozygous loci is found by setting the rate of conversion to heterozygotes, Equation 6.1, equal to the rate of conversion away from heterozygotes, Equation 6.2, and solving for n,

$$\hat{n} = \frac{2Lh(1+hs)}{1-h},$$

which is Equation 1 in the Kirkpatrick and Jenkins paper.

In sexual species, the A_1 allele sweeps through the population without waiting for a second mutation. The contrast in the dynamics of sexual and asexual species is illustrated in Figure 6.3. The asexual species is at a relative disadvantage while sitting on the plateau waiting for the second mutation to sweep through.

The fitness contribution of each of the $\hat{n}$ heterozygous loci to the parthenogenetic individual's overall fitness is $1+hs$. If the overall fitness of an individual is the product of the fitnesses of individual loci (multiplicative epistasis), then the total fitness contribution of the $\hat{n}$ heterozygous loci is $(1+hs)^{\hat{n}}$. These same loci in the sexual species will have sped to fixation, so their total fitness contribution is $(1+s)^{\hat{n}}$. The relative advantage of the sexual species is

$$W_s = \left(\frac{1+s}{1+hs}\right)^{\hat{n}},$$

which is Equation 2 in the Kirkpatrick and Jenkins paper.

Suppose, for example, that $s = 0.01$, $L = 100$, $h = 1/2$; then the number of heterozygous loci is $\hat{n} = 201$ and the relative advantage of the sexual species is $W_s = 2.7$, which is sufficient to overcome the twofold cost of sex.

Problem 6.1 *Selection of 1 percent might be considered unrealistically large. How many loci must be involved to overcome the twofold cost of sex when* $s = 0.001$ *and* $h = 1/2$*?*

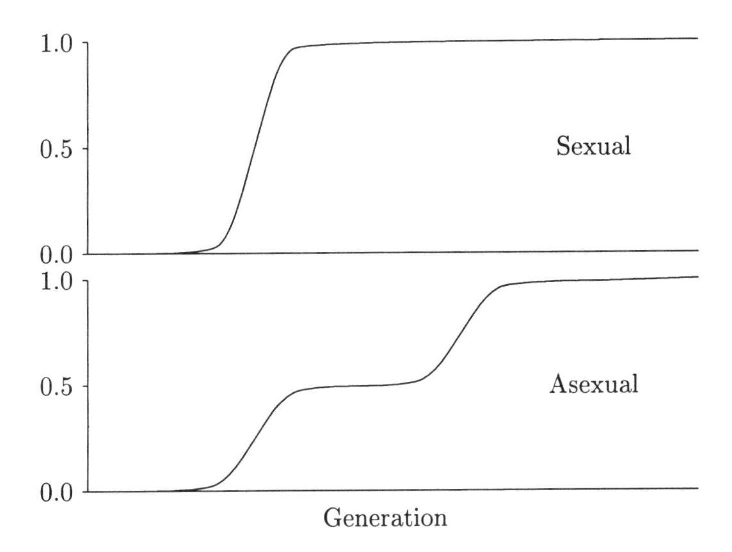

Figure 6.3: The trajectory of allele frequencies under directional selection in asexual and sexual species.

If a species is constantly evolving, then sex with its accompanying segregation will speed up the fixation of advantageous mutations and thereby raise the mean fitness of the population. We have no idea about plausible values of L and s for natural populations, so we cannot say with any confidence that sex is maintained due to the benefits of segregation. We can say with confidence that segregation could explain the maintenance of sex.

There are some reservations about Kirkpatrick and Jenkins' model, though not about the basic idea. Foremost among these is the assumption that the total fitness of an individual is obtained by multiplying the fitness contributions of individual loci. The experimental evidence, as will be shown later, suggests that fitness drops off faster than this as loci interact, the quantitative effects of which will be complex, affecting both $\hat{n}$ and W_s but not affecting the fundamental conclusion that segregation speeds up the rate of substitution of incompletely dominant mutations. Other reservations may be found in the paper itself.

6.2 Crossing-over

In addition to segregation, sex is also associated with crossing-over and its consequence, recombination. As an evolutionary force, recombination causes alleles at different loci eventually to become randomly associated with one another. There is an obvious evolutionary advantage to recombination if there are frequent situations where a species' fitness will be enhanced by rapidly bringing together particular alleles onto the same chromosome. For example, suppose a species is fixed for the A_2 allele at the A locus and the B_2 allele at the closely linked B locus and that the environment suddenly changes such that the A_1

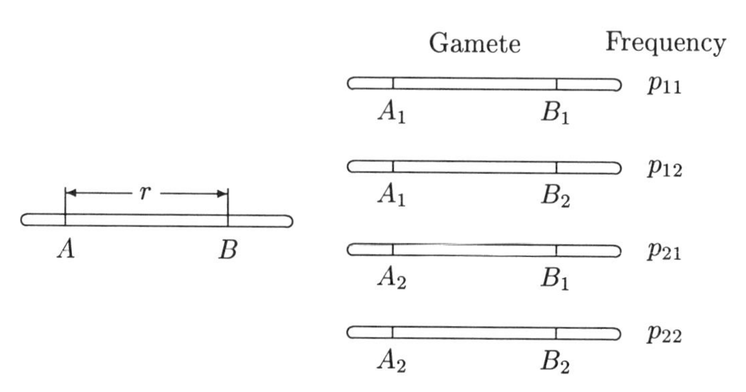

Figure 6.4: The chromosome on the left shows the position of the A and B loci. The right side illustrates the four possible gametes with their frequencies.

and B_1 alleles are now favored. If there is no recombination, then evolution will most likely proceed by one of two routes. Either a mutation to A_1 will appear and the A_1 allele will increase in frequency, followed by the appearance and increase of a B_1 mutation on an A_1-containing chromosome, or the B_1 mutation will appear first, followed by an A_1 mutation on a B_1-containing chromosome. (Or, rarely, both events could happen.) The important points are that two mutational events are required in the same lineage in order for the population to achieve fixation at both loci and that the second mutational event is unlikely to happen until the frequency of the chromosomes carrying the first mutation is very high.

If recombination is allowed and if both the A_1 and B_1 mutations appear and increase, recombination can cause the two to come together on the same chromosome without requiring a second mutation on the same lineage. Moreover, this can occur before either A_1 or B_1 mutations are very frequent. Like segregation, recombination appears to speed up evolution. In the case of recombination, the speedup is due to the production of a recombinant gamete that is more fit than other gametes in the population.

To appreciate some of the subtleties of this argument, which was published in 1932 by the geneticist H. J. Muller, we need to learn more about the evolutionary consequences of recombination. The simplest model capable of showing the effects of recombination is of a diploid species with two linked loci, each with two segregating alleles. The left-hand side of Figure 6.4 illustrates the position of the two loci on the chromosome. The probability that a recombinant gamete is produced at meiosis is denoted by r, which is often called the recombination rate. (The genetic or map distance between the loci is always greater than r because it is the average number of recombinational events rather than the probability of producing a recombinant offspring.)

The right-hand side of Figure 6.4 shows that there are four gametes in the population, A_1B_1, A_1B_2, A_2B_1, and A_2B_2 with frequencies p_{11}, p_{12}, p_{21}, and p_{22}, respectively. The frequency of the A_1 allele, as a function of the gamete frequencies, is $p_{1\cdot} = p_{11} + p_{12}$. The dot in the subscript acts as a placeholder to remind you that there is another locus around, but that we are ignoring it. Similarly, the allele frequency of B_1 is $p_{\cdot 1} = p_{11} + p_{21}$.

Recombination changes the frequencies of these gametes in a very simple way. For example, the frequency of the A_1B_1 gamete after a round of random mating, p'_{11}, is simply

$$p'_{11} = (1-r)p_{11} + rp_{1\cdot}p_{\cdot 1}\,. \tag{6.3}$$

This expression is best understood as a statement about the probability of choosing an A_1B_1 gamete from the population. A randomly chosen gamete will have had one of two possible histories: Either it will be a recombinant gamete (this occurs with probability r) or it won't be (this occurs with probability $1-r$). If it is not a recombinant, then the probability that it is an A_1B_1 gamete is p_{11}. Thus, the probability that the chosen gamete is an unrecombined A_1B_1 gamete is $(1-r)p_{11}$, which is the first term on the right side of Equation 6.3. If the gamete is a recombinant, then the probability that it is A_1B_1 is the probability that the A locus is A_1, which is just the frequency of A_1, $p_{1\cdot}$, times the probability that the B locus is B_1, $p_{\cdot 1}$. The probability of being a recombinant gamete and being A_1B_1 is $rp_{1\cdot}p_{\cdot 1}$, which is the rightmost term of Equation 6.3. The allele frequencies can be multiplied because the effect of a recombination is to choose the allele at the A and B loci independently.

Problem 6.2 *Derive the three equations for the frequencies of the A_1B_2, A_2B_1, and A_2B_2 gametes after a round of random mating.*

The change in the frequency of the A_1B_1 gamete in a single generation of random mating is, from Equation 6.3,

$$\Delta_r p_{11} = r(p_{1\cdot}p_{\cdot 1} - p_{11}). \tag{6.4}$$

The equilibrium gamete frequency is obtained by solving $\Delta_r p_{11} = 0$,

$$\hat{p}_{11} = p_{1\cdot}p_{\cdot 1}.$$

If there were no tendency for the A_1 allele to be associated with the B_1 allele, the probability of choosing an A_1B_1 allele from the population would be the product of the frequencies of the A_1 and B_1 alleles. As this is precisely the equilibrium state of the population, we conclude that recombination removes associations between alleles on chromosomes. The rate of change of the gamete frequency due to recombination is simply the recombination rate, r.

The deviation of the frequency of A_1B_1 from its equilibrium value is called linkage disequilibrium,

$$D = p_{11} - p_{1\cdot}p_{\cdot 1}\,. \tag{6.5}$$

Thus, the frequency of the A_1B_1 gamete may be written

$$p_{11} = p_{1\cdot}p_{\cdot 1} + D,$$

which emphasizes that the departure of the gamete frequency from its equilibrium value is determined by D. Obviously, at equilibrium $D = 0$. The linkage disequilibrium may also be written in the more conventional form

$$D = p_{11}p_{22} - p_{12}p_{21}, \tag{6.6}$$

which leads to the following new expressions for the gamete frequencies:

Gamete:	A_1B_1	A_1B_2	A_2B_1	A_2B_2
Frequency:	p_{11}	p_{12}	p_{21}	p_{22}
Frequency:	$p_{1\cdot}p_{\cdot 1} + D$	$p_{1\cdot}p_{\cdot 2} - D$	$p_{2\cdot}p_{\cdot 1} - D$	$p_{2\cdot}p_{\cdot 2} + D$

Problem 6.3 *Show that the gamete frequencies as a function of D are correct in the above table. Next, show that the definition of D in Equation 6.6 is consistent with the expressions for gamete frequencies by substituting $p_{1\cdot}p_{\cdot 1}+D$ for p_{11} and the equivalent expressions for the other gametes into right side of Equation 6.6, proceeding with a feeding frenzy of cancelations, and ending with a lone D.*

The A_1B_1 and A_2B_2 gametes are often called coupling gametes because the same subscript is used for both alleles. The A_1B_2 and A_2B_1 gametes are called repulsion gametes. Linkage disequilibrium may be thought of as a measure of the excess of coupling over repulsion gametes. When D is positive, there are more coupling gametes than expected at equilibrium; when negative, there are more repulsion gametes than expected.

The value of D after a round of random mating may be obtained directly from Equation 6.3 by using $p_{11} = p_{1\cdot}p_{\cdot 1} + D$,

$$p'_{1\cdot}p'_{\cdot 1} + D' = (1-r)(p_{1\cdot}p_{\cdot 1} + D). + rp_{1\cdot}p_{\cdot 1}.$$

A few quick cancellations yield

$$D' = (1-r)D.$$

Some of the cancellations use the Hardy-Weinberg truism that allele frequencies don't change with random mating. We would be in trouble if the addition of loci affected the Hardy-Weinberg law for single loci!

The change in D in a single generation is

$$\Delta_r D = -rD,$$

which depends on the gamete frequencies only through their contributions to D. Finally,

$$D_t = (1-r)^t D_0,$$

showing, once again, that the ultimate state of the population is $D = 0$. Note that with free recombination ($r = 1/2$) the linkage disequilibrium does not disappear in a single generation. If you find this startling, follow D for a couple of generations in a population initiated with $p_{11} = p_{22} = 1/2$ and $r = 1/2$.

In natural populations, the reduction in the magnitude of linkage disequilibrium by recombination is opposed by several evolutionary forces that may increase $|D|$. Natural selection will increase D if selection favors coupling gametes over repulsion gametes or decrease D if repulsion gametes are favored. Migration may increase the absolute value of D if the allele frequencies of the immigrants differ from those of the resident population. Finally, genetic drift can lead to changes in D due to random sampling. As recombination is most effective for loosely linked loci, we would expect to find tightly linked loci farther from linkage equilibrium than loosely linked loci. This has been seen in molecular data, but the linked loci must be very close indeed in order to see significant levels of disequilibrium.

Returning to Muller's argument that recombination should speed up evolution, our newfound knowledge of the evolutionary consequences of recombination has turned up a potential problem. Recombination not only brings alleles together, but also breaks down associations between alleles. This can actually retard the rate of evolution if the repulsion gametes, A_1B_2 and A_2B_1, are less fit than the A_1B_1 gamete. For an in-depth discussion of Muller's model and other aspects of the evolution of sex, read John Maynard Smith's *Evolutionary Genetics* (1989).

6.3 Muller's ratchet

One advantage of sex is to speed up evolution through either segregation or recombination. In a world populated by evolving predators, prey, and pathogens and in a physical environment that is constantly changing, the ability to evolve quickly must enhance a species' long-term fitness appreciably. The origin and maintenance of sex may be attributed to its role in speeding up the rate of evolution. In addition, sex also plays a role in removing deleterious mutations from a population. H. J. Muller was the first to point this out by describing a phenomenon that has come to be called Muller's ratchet. Muller pointed out that, without sex, deleterious mutations may accumulate on chromosomes faster than selection can remove them from the population, leading to the ultimate extinction of the species. The operation of the ratchet involves three evolutionary forces: mutation, selection, and genetic drift, as will be described in this section.

Picture a population of N parthenogenetic diploid individuals. Each individual will have a certain number of deleterious mutations sprinkled around on its chromosomes. Assume that the mutation rate to deleterious alleles at any particular locus is so small that all deleterious mutations are heterozygous with the normal allele. The individuals may be grouped according to the number of deleterious mutations that they possess. A fraction x_0 of individuals will have no deleterious mutations, a fraction x_1 will have one deleterious mutation, and so forth. The left side of Figure 6.5 shows these classes of individuals. Note that the class of individuals with i deleterious mutations is genetically heterogeneous because different individuals are likely to have their i deleterious mutations at

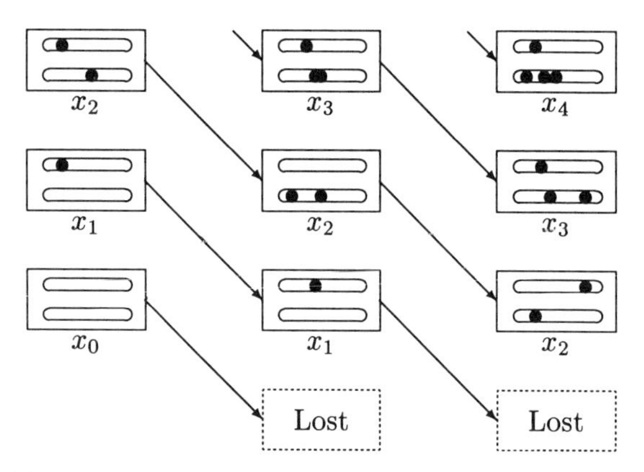

Figure 6.5: An illustration of two clicks of Muller's ratchet. The boxes represent the class of individuals with i deleterious mutations. The two chromosomes within each box represent typical individuals. The number of generations between each column of boxes depends on the efficacy of genetic drift.

different loci.

If the number of individuals with no mutations, Nx_0, is small, then this class will be subject to the action of genetic drift, just as a rare mutation is subject to genetic drift. (Demographic stochasticity is the only source of randomness, as segregation does not occur in asexual organisms.) Should drift cause the loss of all individuals with no mutations, then each individual in the population will have at least one deleterious mutation, and Muller's ratchet will have clicked once. This is illustrated in Figure 6.5 by the shift from column 1 to column 2. If the number of individuals with one mutation, Nx_1, is small, then this class will be subject to loss by genetic drift. If this class is lost, each individual will have at least two mutations, and the ratchet will have clicked once again. The rate of clicking of the ratchet is set by the time to loss of the smallest class by genetic drift. If the parameters are appropriate, then the species slowly accumulates deleterious mutations, leading to its eventual extinction.

For Muller's ratchet to work, the numbers of individuals in the class with the fewest mutations must be small. Population genetics can tell us the size of this class, if we are willing to make a few critical assumptions. The most important assumption is that the fitness of an individual with i deleterious mutations is

$$w_i = (1 - hs)^i.$$

The fitness contributions of the i heterozygous loci for deleterious mutations are multiplied together to obtain the fitness of the individual. The interaction of alleles at different loci is called epistasis; the form of epistasis used here is called multiplicative epistasis. Multiplicative epistasis implies that the effects on the probability of survival of individual loci are independent. If your chance of surviving from the effects of each of two loci in isolation is 1/2, then your chance of surviving their joint effects is 1/4. Later in this chapter we will consider another form of epistasis and some of the relevant experimental literature.

The derivation of the frequency of the classes of individuals with i mutations requires a knowledge of the Poisson distribution. Those unfamiliar with this distribution should read the description on page 159. Our basic strategy is to assume that the number of deleterious mutations in an individual is Poisson-distributed and then to find the mean of the Poisson as a function of the mutation rate and the strength of selection. Why a Poisson? You will have to wait until the derivation is finished; then you will see.

Assume, then, that the number of mutations per individual is Poisson-distributed with mean μ_K,

$$\text{Prob}\{i \text{ mutations}\} = \frac{e^{-\mu_K}\mu_K^i}{i!}.$$

With each reproduction, assume further that an offspring receives a Poisson-distributed number of new mutations with mean U,

$$\text{Prob}\{j \text{ new mutations}\} = \frac{e^{-U}U^j}{j!}.$$

The number of mutations after reproduction is the sum of two Poisson distributions, one with mean μ_K representing the number of mutations per individual before reproduction and one with mean U representing the number of new mutations. As the sum of two Poisson random variables is Poisson-distributed with mean equal to the sum of the means of the two Poissons, we see immediately that the mean number of mutations per individual after reproduction is Poisson with mean

$$\mu' = \mu_K + U.$$

Thus, the frequency of individuals with i mutations after reproduction is

$$x_i' = \frac{e^{-\mu'}\mu'^i}{i!}.$$

The frequency of an individual with i mutations after selection is proportional to its frequency before selection times its fitness,

$$x_i'' = \frac{x_i'(1-hs)^i}{\sum_{j=0}^{\infty} x_j'(1-hs)^j}. \tag{6.7}$$

The numerator is

$$\frac{e^{-\mu'}[\mu'(1-hs)]^i}{i!}.$$

The denominator is

$$\begin{aligned}\sum_{j=0}^{\infty} x_j'(1-hs)^j &= e^{-\mu'}\sum_{j=0}^{\infty}\frac{[\mu'(1-hs)]^j}{j!} \\ &= e^{-\mu'}e^{\mu'(1-hs)}.\end{aligned}$$

The third line is obtained from the Taylor series expansion of the exponential function,

$$e^x = \sum_{i=0}^{\infty} \frac{x^i}{i!}.$$

Thus, Equation 6.7 becomes

$$\frac{e^{-\mu'(1-hs)}[\mu'(1-hs)]^i}{i!},$$

which is, once again, a Poisson distribution. The mean of the new Poisson distribution is

$$\mu'_K = (\mu_K + U)(1 - hs).$$

At equilibrium, the mean does not change, so

$$\hat{\mu}_K = \frac{U(1-hs)}{hs} \approx \frac{U}{hs},$$

from which we obtain the frequency of individuals with i mutations at equilibrium,

$$\hat{x}_i \approx \frac{e^{-U/hs}(U/hs)^i}{i!}.$$

For Muller's ratchet to operate, the numbers of individuals with no mutations must be small enough that they will be eliminated in a reasonable time by genetic drift. Suppose, for example, that the total genomic mutation rate to deleterious alleles is $U = 1$, that the average selection against mutations in the heterozygous state is $hs = 0.1$, and that the population size is $N = 10^5$. In this case the number of individuals with no deleterious mutations is

$$Nx_0 = 4.54,$$

which is so small that genetic drift will soon eliminate them, even though they are the most fit genotypes in the population. However, if the population size is much larger, say $N = 10^7$, then $Nx_0 = 454$ and the ratchet comes to a grinding halt because the waiting time to fix the class of individuals with the fewest mutations becomes extraordinarily long. In those cases where the ratchet does operate, the accumulation of deleterious mutations proceeds at a rate determined by N, hs, and U. As there is no sex, hence no recombination, there is no way to reverse the steady reduction of fitness.

Problem 6.4 *Ten percent selection per locus is too high. Heterozygotes for lethals, for example, usually are at about a 2 to 5 percent disadvantage in* Drosophila. *If hs is changed to 0.02, what happens?*

When there is sex, recombination can generate chromosomes with fewer (and more) mutations, as seen in Figure 6.6. Thus, the classes with fewer mutations are continually regenerated so that sex, with its accompanying recombination,

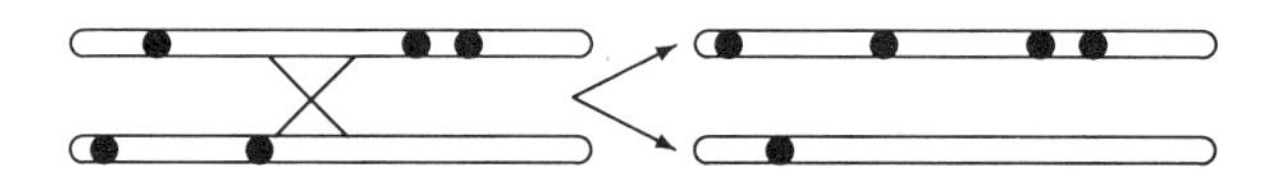

Figure 6.6: An illustration of the production of a chromosome with fewer deleterious mutations than in either parental chromosome because of crossing-over. The filled circles represent deleterious mutations.

may slow down or even reverse the clicking of Muller's ratchet. For this reason, sex is viewed as a way of eliminating deleterious mutations and hence is favored by natural selection. Unfortunately, the story must end here because very little is known about the dynamics of deleterious mutations in finite populations with recombination and even less is known about the evolution of mutations promoting sex.

Because of the difficulty in obtaining a quantitative value for its advantage, it remains unclear whether Muller's ratchet can account for the maintenance of sex. However, the ratchet is often evoked with confidence to explain the loss of genes from chromosomes that do not recombine. For example, in many species in which the male is the heterogametic sex, the Y chromosome does not recombine with either the X chromosome or other Y chromosomes. Without recombination, the Y chromosome should slowly accumulate deleterious mutations and, quite plausibly, eventually fail to have any functional genes at all. In many species, including our own, the Y chromosome is almost entirely heterochromatic with only a few loci. Muller's ratchet is a perfectly viable explanation for this fact.

6.4 Kondrashov's hatchet

The assumption of multiplicative epistasis in the model of Muller's ratchet is contradicted by a number of experimental studies. One such study by Terumi Mukai (1968) is illustrated in Figure 6.7. In this study, Mukai accumulated mutations on second chromosomes of *Drosophila melanogaster* for 60 generations under conditions that minimized natural selection. As the mutations accumulated, the relative viability of the second chromosomes when homozygous decreased. Mukai estimated that the mutation rate to deleterious mutations with measurable effects was 0.1411 per second chromosome per generation. For example, at 60 generations, a typical second chromosome would have $60 \times 0.1411 = 8.46$ mutations, which corresponds to the rightmost point on the graph.

The three curves on the graph correspond to three different models of epistasis. The lower convex curve is for multiplicative epistasis of the form

$$w_n = (1 - s)^n,$$

where n is the number of deleterious mutations, s is a measure of the effect of each mutation, and w_n is the relative viability of a fly with n homozygous

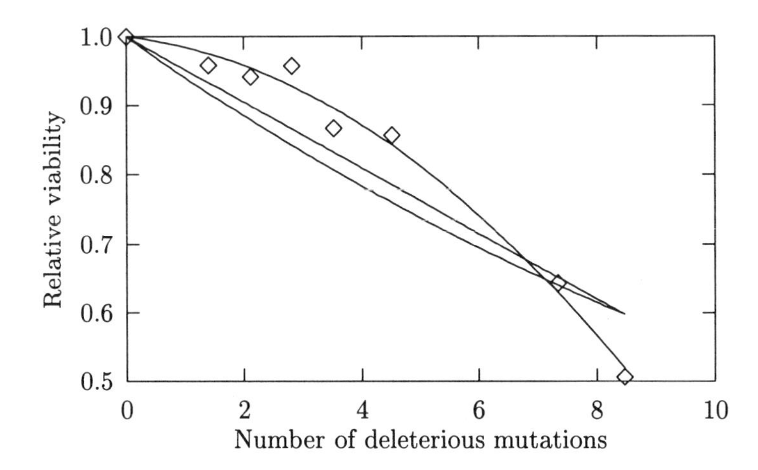

Figure 6.7: The relative viability as a function of the inferred number of homozygous deleterious mutations in *Drosophila melanogaster.* The upper concave curve is a quadratic synergistic model, the middle straight line is an additive model, and the lower convex curve corresponds to multiplicative epistasis. The data are from Mukai (1968).

deleterious mutations. The method of least squares was used to estimate s, whose value is 0.05897. The middle straight line corresponds to additive epistasis

$$w_n = 1 - sn$$

with $s = 0.04749$. Finally, the upper concave curve is for a quadratic model of synergistic epistasis,

$$w_n = 1 - sn - an^2,$$

where $s = 0.009813$ and $a = 0.00555$. The figure leaves little doubt that synergistic epistasis fits the data much better than do additive or multiplicative epistasis. With synergistic epistasis, fitness drops off faster than expected from the effects of mutations in isolation. It is an embodiment of the notion of "things going from bad to worse."

Synergistic epistasis does not invalidate Muller's ratchet, but it does change its rate. Unfortunately, the multiplicative assumption is directly responsible for the Poisson distribution of the number of mutations per genome. Changing this assumption necessitates a much more difficult mathematical development. Rather than pursuing that gruesome prospect, a much more exciting direction is to investigate the effects of synergistic epistasis itself on the fate of sexual and asexual species. This will lead to another argument for the evolutionary advantage of sex based on the ability of sexual species to eliminate deleterious mutations more effectively than do asexual species.

The importance of synergistic epistasis for the maintenance of sex was emphasized in a paper by Alexey Kondrashov (1988), whose model was dubbed Kondrashov's hatchet by Michael Turelli, who is never wanting for a clever turn

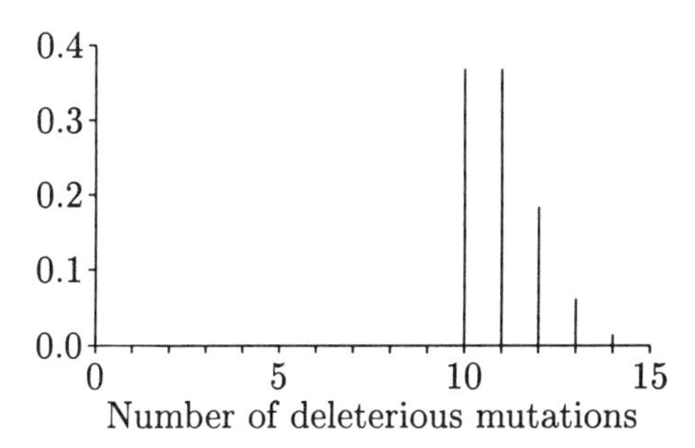

Figure 6.8: The frequency of asexual individuals just after reproduction and just before selection when $k = 10$ and $U = 1$.

of phrase. Models of synergistic epistasis are very difficult to analyze. However, Kondrashov realized that truncation selection, which is well understood by quantitative geneticists, is really an instance of synergistic epistasis. Imagine that "deleterious" mutations accumulate in a genome with no effect on fitness until the number of mutations exceeds some magic number, k, at which point any additional mutations are lethal. Said another way: Fitness, as a function of the number of deleterious mutations, is equal to one until the number of mutations is greater than k; when the number of deleterious mutations is greater than k, fitness is zero. The fitness could be written as

$$w_n = \begin{cases} 1 & \text{if } n \leq k \\ 0 & \text{if } n > k \end{cases} .$$

The hatchet falls on all genotypes with more than k mutations.

If the number of mutations is now viewed as a quantitative trait, then Kondrashov's hatchet is formally the same as truncation selection, as studied in Chapter 5. Accordingly, we will use our results on the response to selection to study the evolution of the number of deleterious mutations per individual in sexual organisms. But first, we will record some observations about selection in asexual organisms to see the relative advantages of being sexual.

Mutation is added to Kondrashov's model in the same way as was done for Muller's ratchet. Each generation, every offspring receives a Poisson-distributed number of new deleterious mutations. The mean of the Poisson is U. Through time, the number of deleterious mutations in any lineage will increase until it reaches k. From then on, any offspring with additional mutations will die. After a sufficient period of time, every individual in the population will have exactly k mutations. From this point on, the frequency of offspring in the population with no additional mutations is the probability that a Poisson random variable with mean U is zero,

$$\text{Prob}\{\mathbf{X} = 0\} = e^{-U}.$$

The probability that an offspring has one or more new mutations is $1 - e^{-U}$. The frequency of individuals with $k = 10$ or more mutations, before selection and when $U = 1$, is illustrated in Figure 6.8.

The genetic load of the equilibrium asexual population is just

$$\frac{w_{\max} - \bar{w}}{w_{\max}} = 1 - [e^{-U} \times 1 + (1 - e^{-U}) \times 0]$$
$$= 1 - e^{-U}.$$

Once again, we have a calculation that shows that the genetic load depends only on the mutation rate and is independent of the strength of selection. For small U, the load is approximately

$$1 - (1 - U) = U,$$

which is the same as the result we obtained on page 62. There, the contribution of a single locus to the load was found to be $2u$, where u is the mutation rate per allele. Here, U is the mutation rate per diploid genome, which is twice the rate per haploid genome. This difference accounts for the factor of two.

As an aside, it is worth noting that the equilibrium mutational load for an asexual species with any sort of epistasis is remarkably easy to derive by considering only the frequency of the genomes that are free of deleterious mutations, x_0. In the next generation, the frequency of this class is

$$x_0' = \frac{x_0 w_0 e^{-U}}{\bar{w}}.$$

That is, its frequency is equal to its frequency in the current generation times the probability that an offspring gets no additional mutations (e^{-U} is the probability that the Poisson random variable is zero) times the fitness of the deleterious-free class, w_0, which is one in our models, divided by the mean fitness of the population. At equilibrium, $x_0' = x_0$, so

$$\bar{w} = e^{-U}.$$

The simplicity of the result might be unsettling at first, but really is as simple as it appears.

Consider now the situation in a sexual species. As a first step, imagine what would happen if the asexual species were suddenly to turn sexual. With meiosis and crossing over, individuals would appear with fewer than k mutations. They, along with those with k mutations, would live, while those with more than k would die. As a result, the mean number of deleterious alleles would be less than in the asexual population. With sex, selection lowers the mean number of mutations per individual, while mutation increases the number. Eventually an equilibrium is reached where the genetic load must be lower than the asexual load. By how much lower? Read on!

Kondrashov's hatchet kills off all offspring with more than k mutations, which reduces the average number of deleterious mutations per individual by an amount that is, by definition, the selection differential, S. As there is no environmental contribution to the number of deleterious mutations, the heritability of this phenotype is $h^2 = 1$. By Equation 5.11, the response to selection is

$$R = h^2 S = S.$$

Thus, each generation the average number of mutations per individual is increased by U and decreased by $R = S$. Eventually, an equilibrium will be reached where

$$U = S. \tag{6.8}$$

Our job is to discover the mutational load at this equilibrium and to see if it is sufficiently smaller than the asexual load to overcome the twofold cost of sex.

Recall from Chapter 5 that the selection intensity is

$$i(p) = \frac{S}{\sqrt{V_P}}.$$

Dividing both sides of Equation 6.8 by $\sqrt{V_P}$, we find that the selection intensity at equilibrium is

$$i(p) = \frac{U}{\sqrt{V_P}}.$$

The rightmost term, the ratio of the mean number of deleterious mutations in an offspring divided by the standard deviation of the total number of deleterious mutations per individual, is subject to experimental investigation. In *Drosophila*, Kondrashov argued that $U = 2$ and $\sqrt{V_P} = 10$, in which case the intensity of selection is $i(p) = 0.2$. Referring to Figure 5.6, we see that this intensity occurs when about 10 percent of the population is killed each generation by Kondrashov's hatchet. This is the mutational load of a sexual population at equilibrium. The genetic load of an asexual species is $1 - e^{-U} = 0.86$, much greater than the genetic load of sexual species (0.1). The ratio of the sexual to asexual mean fitnesses is about $0.9/0.14 = 6.4$, which easily overcomes the twofold cost of sex.

In this chapter, we have examined four possible explanations for the maintenance of sex in face of its twofold cost. Two of the explanations, those of Kirkpatrick and Jenkins and of Kondrashov, can be developed to the point of showing quantitatively that sex can plausibly achieve a twofold advantage. We have not addressed the actual evolution of sex. That is, we have not followed the fate of a mutation that confers sexuality on its bearer to see that sex will, in fact, evolve. Models of the evolution of sex are very difficult to analyze.

6.5 Answers to problems

6.1 The number of heterozygous loci needed to overcome the twofold cost satisfies

$$(1.001/1.0005)^{\hat{n}} = 2$$

which gives $\hat{n} = 1387$. The number of loci, L, as a function of $\hat{n}$ is

$$L = \frac{\hat{n}(1-h)}{2h(1+hs)} = 693.$$

6.2 Derive the three equations for the frequencies of the A_1B_2, A_2B_1, and A_2B_2 gametes after a round of random mating.

$$\begin{aligned} p'_{12} &= (1-r)p_{12} + rp_{1\cdot}p_{\cdot 2} \\ p'_{21} &= (1-r)p_{21} + rp_{2\cdot}p_{\cdot 1} \\ p'_{22} &= (1-r)p_{22} + rp_{2\cdot}p_{\cdot 2} \,. \end{aligned}$$

6.3 The verification for A_1B_2 is

$$\begin{aligned} p_{12} &= p_{1\cdot}p_{\cdot 2} - D \\ &= (p_{11} + p_{12})(p_{12} + p_{22}) - (p_{11}p_{22} - p_{12}p_{21}) \\ &= p_{12}. \end{aligned}$$

The other cases are done similarly. The verification of Equation 6.6 is easy (and fun).

6.4 For $U = 1$ and $hs = 0.02$,

$$e^{U/hs} = 1.9 \times 10^{-22}.$$

The problem, of course, is that selection is so weak that mutation dominates and the simple dynamics of Muller's ratchet break down. (Unless you feel that population sizes are larger than 10^{22}.)

Appendix A

Mathematical Necessities

The mathematics used in this book are mostly those found in high school algebra courses. The exceptions are certain approximations based on Taylor series expansions, which will be summarized in this appendix, and some calculus that appears mostly in the advanced topics at the ends of some chapters.

The Taylor series expansion of the function $f(x)$ around x_0 is

$$f(x) = \sum_{n=0}^{\infty} \frac{f^{(n)}}{n!}(x - x_0)^n,$$

where $f^{(n)}$ is the nth derivative of f. In population genetics, satisfactory approximations to a function are often obtained by using only the first three terms in the expansion,

$$f(x) \approx f(x_0) + f'(x_0)(x - x_0) + \frac{1}{2}f''(x - x_o)^2. \tag{A.1}$$

The most commonly employed approximation in population genetics is of a product of numbers, each close to one in magnitude. For example, suppose we wish to approximate the product

$$\prod_{i=1}^{n}(1 + x_i), \tag{A.2}$$

where $|x_i|$ is close to zero. When faced with a product, your first thought should always be to convert it to a sum. Sums are often easier to work with, particularly in the present context. The conversion is done with the natural logarithm, as this function has the property that the log of a product is the sum of the logs, $\ln(ab) = \ln(a) + \ln(b)$. However, we can't simply take the log of Equation A.2, as this would change its value. Rather, we take the exponential* of the log, as the

*The exponential function is written sometimes as e^x and sometimes as $\exp(x)$.

exponential and logarithm functions are inverses of each other, $\exp(\ln(x)) = x$. Taking the exponential of the log of Equation A.2 gives

$$e^{\sum_{i=1}^{n} \ln(1+x_i)}.$$

At this point we need the first of our function approximations,

$$\ln(1+x) \approx x - x^2/2 \tag{A.3}$$

when x is close to zero. If $|x|$ is very small, say $x = 0.1$, then we may usually ignore the squared term, as its value is only 0.5 percent of the value of the first term, x. Applying this approximation to our product gives

$$\prod_{i=1}^{n}(1+x_i) \approx e^{\sum_{i=1}^{n} x_i}, \tag{A.4}$$

which is usually as far as we need go. In some cases, we may want to use the approximation for the exponential function,

$$e^x \approx 1 + x + x^2/2, \tag{A.5}$$

which is valid when x is close to zero, to approximate Equation A.4 with

$$\prod_{i=1}^{n}(1+x_i) \approx 1 + \sum_{i=1}^{n} x_i, \tag{A.6}$$

when x is sufficiently small.

Problem A.1 *Suppose $x_i = 0.1$ and $n = 10$. Calculate the product and the two approximations to the product. Do the same thing for $x = 0.05$. Notice how much the approximations improved in the second case.*

Another approximation that is frequently used in population genetics is based on the sum of a geometric series,

$$1 + x + x^2 + x^3 + \cdots = \frac{1}{1-x},$$

when $|x| < 1$. When x is small,

$$\frac{1}{1-x} \approx 1 + x + x^2. \tag{A.7}$$

Three types of means occur in population genetics: the arithmetic, harmonic, and geometric. The arithmetic mean of n numbers is the familiar

$$a = \frac{x_1 + x_2 + \cdots + x_n}{n}.$$

The geometric mean is the nth root of the product,

$$g = (x_1 x_2 \ldots x_n)^{1/n},$$

where $x_i \geq 0$. Finally, the harmonic mean is the reciprocal of the arithmetic mean of the reciprocals,

$$h = \left(\frac{1/x_1 + 1/x_2 + \cdots + 1/x_n}{n} \right)^{-1},$$

where $x_i > 0$. A famous inequality from classical mathematics is

$$a \geq g \geq h,$$

where equality holds only when all of the x_i are equal.

Appendix B

Probability

Many of the main ideas in population genetics involve some element of randomness. Genetic drift is the prime example, but even as seemingly a nonrandom quantity as the mean fitness of the population, $\bar{w}$, is couched in the vocabulary of probabilities. Population genetics uses only the most basic elements of probability theory, but these elements are crucial to a true understanding of the field. This appendix contains everything that is required. It is not meant to substitute for a proper probability course, but it may serve as a reminder of things learned elsewhere or, for some, a telegraphic but complete background for the book.

Probability theory is concerned with the description of experiments whose outcomes cannot be known with certainty. Rather, a certain probability is associated with each outcome. In population genetics, we are usually interested in attaching some numerical value to the outcome of an experiment. The value may be the frequency of an allele, the height of an individual, the fitness of a genotype, or some other quantity. We are frequently interested in the mean or variance of these values. Random variables are the constructs that capture the notion of numerically valued outcomes of an experiment. Thus, this appendix is mostly about random variables.

What are random variables?

A discrete random variable, for example $\mathbf{X}$, is a function that takes on certain values depending on the outcome of some event, trial, or experiment. The various outcomes have probabilities of occurring; hence, the values of the random variable have probabilities of occurring. An event with n outcomes and its associated random variable may be described as follows:

Outcome	Value of $\mathbf{X}$	Probability
1	x_1	p_1
2	x_2	p_2
$\vdots$	$\vdots$	$\vdots$
i	x_i	p_i
$\vdots$	$\vdots$	$\vdots$
n	x_n	p_n

To be sure there is always an outcome, $\sum_{i=1}^{n} p_i = 1$. We say that the probability that the random variable $\mathbf{X}$ takes on the value x_i is p_i, or, more concisely,

$$\text{Prob}\{\mathbf{X} = x_i\} = p_i.$$

The probabilities $p_1, p_2, \ldots$ are often referred to as the probability density or probability distribution of the random variable $\mathbf{X}$.

For example, if we flip a fair coin and attach the value one to the outcome heads and zero to tails, the table becomes:

Outcome	Value of $\mathbf{X}$	Probability
Heads	1	1/2
Tails	0	1/2

Moments of random variables

Two properties of random variables are useful in applications, the mean and the variance. The mean is defined as

$$E\{\mathbf{X}\} = \sum_{i=1}^{n} p_i x_i, \tag{B.1}$$

where E means "expectation of." Notice that the mean is a weighted average of the values taken by the random variable. Those values that are more probable (have a larger p_i) contribute relatively more to the mean than do those values that are less probable. The mean is often denoted by μ.

The variance of a random variable is the expectation of the squared deviations from the mean,

$$\begin{aligned} \text{Var}\{\mathbf{X}\} &= E\{(\mathbf{X} - \mu)^2\} \\ &= \sum_{i=1}^{n} p_i (x_i - \mu)^2 \\ &= \sum_{i=1}^{n} p_i x_i^2 - \mu^2. \end{aligned}$$

A useful observation gleaned from the last line is

$$\text{Var}\{\mathbf{X}\} = E\{\mathbf{X}^2\} - E\{\mathbf{X}\}^2. \tag{B.2}$$

The variance is often denoted by σ^2. The mean is called a measure of central tendency; the variance, a measure of dispersion.

For our coin-flipping example, the mean is

$$\mu = \frac{1}{2} \times 1 + \frac{1}{2} \times 0 = \frac{1}{2},$$

and the variance is

$$\sigma^2 = \frac{1}{2}\left(1 - \frac{1}{2}\right)^2 + \frac{1}{2}\left(0 - \frac{1}{2}\right)^2 = \frac{1}{4}.$$

In population genetics, an important quantity is the mean fitness of the population, $\bar{w}$. The mean fitness does have a proper probabilistic interpretation if we construct a random variable whose values are the fitnesses of genotypes and whose probabilities are the frequencies of genotypes.

Outcome	Value of **X**	Probability
A_1A_1	1	p^2
A_1A_2	$1 - hs$	$2pq$
A_2A_2	$1 - s$	q^2

The mean fitness of the population is

$$\begin{aligned} \bar{w} &= p^2 \times 1 + 2pq \times (1 - hs) + q^2 \times (1 - s) \\ &= 1 - 2pqhs - q^2 s. \end{aligned}$$

While the invocation of a random variable in this setting may seem contrived, it reflects a duality in the definitions of probabilities that runs deep in probability theory. Often it is more natural to refer to the probability of an outcome; other times it is more natural to refer to the relative frequency of an outcome in an experiment that is repeated many times. The definition of the mean fitness falls into the latter framework.

Noteworthy discrete random variables

Bernoulli random variable

These are very similar to our coin-flipping example except that we allow probabilities other than 1/2:

Outcome	Value of **X**	Probability
Success	1	p
Failure	0	$q = 1 - p$

The mean of a Bernoulli random variable is

$$\mu = 1 \times p + 0 \times q = p,$$

and the variance is

$$\sigma^2 = 1^2 \times p + 0^2 \times q - p^2 = pq.$$

Binomial random variable

These random variables represent the number of successes in n independent trials when the probability of success for any one trial is p. The random variable can take on the values $0, 1, \ldots, n$ with probabilities

$$\text{Prob}\{\mathbf{X} = i\} = \frac{n!}{i!(n-i)!} p^i (1-p)^{n-i}, \tag{B.3}$$

where $n! = n(n-1)(n-2)\cdots(2)(1)$ is "n factorial." For example, the probability of three successes in five trials when the probability of success is 0.2 is

$$\text{Prob}\{\mathbf{X} = 3\} = \frac{5 \times 4 \times 3 \times 2 \times 1}{(3 \times 2 \times 1)(2 \times 1)} \times 0.2^3 \times 0.8^2 = 0.0512.$$

As the binomial distribution plays a special role in population genetics, its derivation is of more than passing interest. Consider first the easier problem of finding the probability of a particular sequence of successes and failures in a given experiment. For example, the probability of a success on the first trial, a failure on the next trial, successes on the next two trials, and finally a failure, is

$$\text{Prob}\{\text{SFSSF}\} = pqppq = p^3 q^2,$$

which is precisely the rightmost term in Equation B.3 for the special case of three successes in five trials. It is a small leap to see that this term is the probability of a particular sequence of i successes and $(n-i)$ failures.

To obtain the probability of three successes, we need only calculate the number of different sequences of three S's and two F's, which is precisely the left hand term of the binomial probability

$$\frac{5!}{3!2!} = 10.$$

Each of the 10 sequences has exactly the same probability of occurring, so the total probability of three successes is 10 times $p^3 q^2$.

The final task is to discover the more general result that the number of sequences with i successes and $n-i$ failures is the binomial coefficient

$$\frac{n!}{i!(n-i)!}.$$

Naturally, there is a trick. First, consider the number of sequences of i successes and $(n-i)$ failures when each success and failure is labeled. That is, suppose we call the three successes in the example S_1, S_2, and S_3 and the two failures, F_1 and F_2. The sequence $S_1F_1S_2S_3F_2$ is now viewed as different from the sequence $S_2F_1S_1S_3F_2$. (Without the labeling, these two are the same sequence.) There are $n!$ distinct, labeled sequences. The easiest way to see this is by noting that there are n differently labeled successes or failures that could appear in the first position of the sequence, $n-1$ that could appear in the second position, $n-3$ in the third, etc., for a total of $n!$ distinct sequences.

But we don't care about the labeling, so we must divide the number of labeled sequences by the number of different sequences of just the successes and just the failures. There are $i!$ differently labeled orderings of the labeled S's and $(n-i)!$ different labelings of the failures. Thus, the total number of unlabeled orderings of i successes and $n-i$ failures is

$$\frac{n!}{i!(n-i)!},$$

as was to be shown. As the probability of any one of these unlabeled sequences is $p^i q^{n-i}$, the total probability of i successes is as given in Equation B.3.

The mean of the binomial distribution is

$$\mu = \sum_{i=0}^{n} i \frac{n!}{i!(n-i)!} p^i q^{n-i} = np.$$

If you love algebra, proving this will be a delight. Otherwise, a simple derivation is given on page 163. The derivation of the variance,

$$\sigma^2 = \sum_{i=0}^{n} (i-np)^2 \frac{n!}{i!(n-i)!} p^i q^{n-i} = npq,$$

may be found on page 163.

Poisson random variable

These random variables can take values $0, 1, \ldots, \infty$ with probabilities

$$\text{Prob}\{\mathbf{X} = i\} = \frac{e^{-\mu}\mu^i}{i!}. \tag{B.4}$$

Poisson random variables are obtained by taking the limit of binomial random variables as $n \to \infty$ and $p \to 0$ with the mean $\mu = np$ remaining fixed. To convince yourself that this is true, set $p = \mu/n$ in Equation B.3 and use the facts that

$$\lim_{n\to\infty} \left(1 - \frac{\mu}{n}\right)^n = e^{-\mu}$$

and

$$\frac{n!}{(n-i)!} \sim n^i$$

to obtain Equation B.4. Poisson random variables are used to describe situations where there many opportunities to succeed (n is large), the probability of success on any one trial is small (p is small), and the outcomes of separate trials are independent.

The mean of the Poisson distribution is

$$\sum_{i=0}^{\infty} i \frac{e^{-\mu}\mu^i}{i!} = \mu.$$

Surprisingly, the variance is equal to μ as well. Both of these moments may be obtained from the binomial moments by setting $p = \mu/n$ and letting $n \to \infty$ while holding μ fixed. Try it, you'll like it!

Poisson random variables are unusual in that the sum of two Poisson random variables is also a Poisson random variable. If $\mathbf{X}$ is Poisson with mean μ_x and $\mathbf{Y}$ is an independent Poisson random variable with mean μ_y, then $\mathbf{X} + \mathbf{Y}$ is Poisson with mean $\mu_x + \mu_y$. The proof is not difficult. Perhaps you can see why it must be true by thinking of a Poisson as a large number of opportunities for rare events to occur.

Geometric random variable

The geometric random variable, which can take on the values $1, 2, \ldots, \infty$, describes the time of the first success in a sequence of independent trials with the probability of success being p and the probability of failure, $q = 1 - p$,

$$\text{Prob}\{\mathbf{X} = i\} = q^{i-1}p. \tag{B.5}$$

The mean of the geometric distribution is $1/p$ and the variance is q/p^2.

Correlated random variables

Suppose we have an experiment with each outcome associated with two random variables, X and Y. Their outcomes may be summarized as follows:

	y_1	y_2	y_3	$\cdots$	marginal
x_1	(x_1, y_1) p_{11}	(x_1, y_2) p_{12}	(x_1, y_3) p_{13}	$\cdots$ $\cdots$	$p_{1\cdot}$
x_2	(x_2, y_1) p_{21}	(x_2, y_2) p_{22}	(x_2, y_3) p_{23}	$\cdots$ $\cdots$	$p_{2\cdot}$
x_3	(x_3, y_1) p_{31}	(x_3, y_2) p_{32}	(x_3, y_3) p_{33}	$\cdots$ $\cdots$	$p_{3\cdot}$
$\vdots$	$\vdots$	$\vdots$	$\vdots$	$\ddots$	$\vdots$
marginal	$p_{\cdot 1}$	$p_{\cdot 2}$	$p_{\cdot 3}$	$\cdots$	

The marginal distribution for $\mathbf{X}$ is

$$\text{Prob}\{\mathbf{X} = x_i\} = p_{i\cdot} = \sum_j p_{ij}.$$

The marginal distribution allows us to write the mean of $\mathbf{X}$ as

$$\mu_x = \sum_i \sum_j p_{ij} x_i = \sum_i p_{i\cdot} x_i. \tag{B.6}$$

The variance of $\mathbf{X}$ and the moments of $\mathbf{Y}$ are obtained in a similar fashion.

The covariance of $\mathbf{X}$ and $\mathbf{Y}$ is defined as

$$\begin{aligned}\mathrm{Cov}\{\mathbf{X},\mathbf{Y}\} &= E\{(\mathbf{X}-\mu_x)(\mathbf{Y}-\mu_y)\} \\ &= \sum_i\sum_j p_{ij}(x_i-\mu_x)(y_j-\mu_y) \\ &= \sum_i\sum_j p_{ij}x_iy_j - \mu_x\mu_y. \end{aligned} \tag{B.7}$$

The covariance is a measure of the tendency of two random variables to vary together. If, for example, $\mathbf{X}$ and $\mathbf{Y}$ tend to be large and small together, then their covariance will be positive. If when $\mathbf{X}$ is large, $\mathbf{Y}$ tends to be small, their covariance will be negative. If the two random variables are independent, their covariance is zero (see below).

The correlation coefficient of $\mathbf{X}$ and $\mathbf{Y}$ is defined to be

$$\rho = \mathrm{Corr}\{\mathbf{X},\mathbf{Y}\} = \frac{\mathrm{Cov}\{\mathbf{X},\mathbf{Y}\}}{\sqrt{\mathrm{Var}\{\mathbf{X}\}\mathrm{Var}\{\mathbf{Y}\}}}. \tag{B.8}$$

The correlation coefficient is always between minus one and one, $-1 \leq \rho \leq 1$.

The random variables $\mathbf{X}$ and $\mathbf{Y}$ are said to be independent if

$$p_{ij} = p_{i\cdot}p_{\cdot j}.$$

If they are independent, then their covariance is zero:

$$\sum_i\sum_j p_{ij}x_iy_j - \mu_x\mu_y = \sum_i p_{i\cdot}x_i \sum_j p_{\cdot j}y_j - \mu_x\mu_y = 0.$$

(Two random variables with zero covariance are not necessarily independent.)

For example, in the generalized Hardy-Weinberg we have

Genotype:	A_1A_1	A_1A_2	A_2A_2
Frequency:	x_{11}	x_{12}	x_{22}

We can imagine the state of each of the two gametes in a zygote as being a (correlated) random variable that equals one if the gamete is A_1 and zero if it is A_2:

	A_1	A_2	marginal
A_1	$(1,1)$	$(1,0)$	p
	x_{11}	$x_{12}/2$	
A_2	$(0,1)$	$(0,0)$	q
	$x_{12}/2$	x_{22}	
marginal	p	q	

The moments are

$$\begin{aligned}\mu &= p \\ \sigma^2 &= pq \\ \mathrm{Cov}\{\mathbf{X}, \mathbf{Y}\} &= x_{11} - p^2 \\ \mathrm{Corr}\{\mathbf{X}, \mathbf{Y}\} &= \frac{x_{11} - p^2}{pq}.\end{aligned}$$

The expression for the correlation is the same as for F; hence F is often called the correlation of uniting gametes.

Operations on random variables

The simplest (nontrivial) operation that can be performed on a random variable is to multiply it by a number and add another number. Let $\mathbf{Y}$ be the transformed random variable $\mathbf{Y} = a\mathbf{X} + b$. The mean of $\mathbf{Y}$ is

$$E\{\mathbf{Y}\} = aE\{\mathbf{X}\} + b. \tag{B.9}$$

The proof is as follows:

$$\begin{aligned}E\{\mathbf{Y}\} &= \sum_i (ax_i + b)p_i \\ &= a\sum_i x_i p_i + b\sum_i p_i \\ &= aE\{\mathbf{X}\} + b.\end{aligned}$$

The variance is

$$\mathrm{Var}\{\mathbf{Y}\} = a^2 \mathrm{Var}\{\mathbf{X}\}, \tag{B.10}$$

which may be obtained using an argument similar to that used for the mean.

The distribution of the sum of random variables is often difficult to calculate. However, the mean and variance of a sum are relatively easy to obtain. Let $\mathbf{Z} = \mathbf{X} + \mathbf{Y}$. The mean of $\mathbf{Z}$ may be derived as follows:

$$\begin{aligned}E\{\mathbf{Z}\} &= \sum_i \sum_j (x_i + y_j)p_{ij} \\ &= \sum_i x_i p_{i\cdot} + \sum_j x_j p_{\cdot j},\end{aligned}$$

from which we conclude, using Equation B.6,

$$E\{\mathbf{X} + \mathbf{Y}\} = E\{\mathbf{X}\} + E\{\mathbf{Y}\}. \tag{B.11}$$

Notice that this result is true no matter what dependence there may be between $\mathbf{X}$ and $\mathbf{Y}$.

Equation B.11 may be used to find the expectation of a binomial random variable. Recall that the binomial random variable represents the number of successes in n independent trials when the probability of success on any particular trial is p. We can write the number of successes as

$$\mathbf{X} = \mathbf{X}_1 + \mathbf{X}_2 + \cdots + \mathbf{X}_n,$$

where $\mathbf{X}_i$ is a Bernoulli random variable that is one if a success occurred on the ith trial and zero otherwise. As the expectation of $\mathbf{X}_i$ is p, we have

$$E\{\mathbf{X}\} = nE\{\mathbf{X}_i\} = np,$$

which is, as claimed earlier, the mean of a binomial distribution.

The variance of a sum is

$$\begin{aligned}\mathrm{Var}\{\mathbf{Z}\} &= E\{(\mathbf{X} + \mathbf{Y} - \mu_x - \mu_y)^2\} \\ &= E\{(\mathbf{X} - \mu_x)^2 + 2(\mathbf{X} - \mu_x)(\mathbf{Y} - \mu_y) + (\mathbf{Y} - \mu_y)^2\},\end{aligned}$$

from which we conclude that

$$\mathrm{Var}\{\mathbf{X} + \mathbf{Y}\} = \mathrm{Var}\{\mathbf{X}\} + 2\mathrm{Cov}\{\mathbf{X}, \mathbf{Y}\} + \mathrm{Var}\{\mathbf{Y}\}. \tag{B.12}$$

Using the same sort of argument, you can show that

$$\mathrm{Var}\{a_1\mathbf{X}_1 + a_2\mathbf{X}_2\} = a_1^2\mathrm{Var}\{\mathbf{X}_1\} + a_2^2\mathrm{Var}\{\mathbf{X}_2\} + 2a_1a_2\mathrm{Cov}\{\mathbf{X}_1, \mathbf{X}_2\} \tag{B.13}$$

and

$$\begin{aligned}\mathrm{Cov}\{a_1\mathbf{X}_1 + a_2\mathbf{X}_2, b_1\mathbf{Y}_1 + b_2\mathbf{Y}_2\} = \\ a_1b_1\mathrm{Cov}\{\mathbf{X}_1, \mathbf{Y}_1\} + a_1b_2\mathrm{Cov}\{\mathbf{X}_1, \mathbf{Y}_2\} \\ + a_2b_1\mathrm{Cov}\{\mathbf{X}_2, \mathbf{Y}_1\} + a_2b_2\mathrm{Cov}\{\mathbf{X}_2, \mathbf{Y}_2\}.\end{aligned} \tag{B.14}$$

An important special case concerns independent random variables, $\mathbf{X}_i$, for which the variance of the sum is the sum of the variances,

$$\mathrm{Var}\left\{\sum_i \mathbf{X}_i\right\} = \sum_i \mathrm{Var}\{\mathbf{X}_i\}.$$

This may be used to show that the variance of the binomial distribution is npq exactly as we did to obtain the mean.

Noteworthy continuous random variables

Normal random variable

Continuous random variables take values over a range of real numbers. For example, normal random variables can take on any value in the interval

$(-\infty, \infty)$. With so many possible values, the probability that a continuous random variable takes on a particular value is zero. Thus, we will never find ourselves writing "Prob$\{\mathbf{X} = x\}$ =" for a continuous random variable. Rather, we will write the probability that the random variable takes on a value in a specified interval. The probability is determined by the probability density function, $f(x)$. Using the probability density function, we can write

$$\text{Prob}\{a < \mathbf{X} < b\} = \int_a^b f(x)dx.$$

If $\mathbf{X}$ takes on values in the interval (α, β), then

$$\int_\alpha^\beta f(x)dx = 1.$$

The mean of a continuous random variable is defined as it is for discrete random variables:

$$\mu = E\{\mathbf{X}\} = \int_\alpha^\beta x f(x)dx.$$

Similarly, the variance is

$$\sigma^2 = \int_\alpha^\beta (x - \mu)^2 f(x)dx.$$

The most important of the continuous random variables is the normal or Gaussian random variable, whose probability density function is

$$f(x) = \frac{1}{\sqrt{2\pi\sigma^2}} \exp\left(-\frac{(x-\mu)^2}{2\sigma^2}\right),$$

where the parameters μ and σ^2 are, in fact, the mean and variance of the distribution, respectively. For example, it is possible to show that

$$\mu = \int_{-\infty}^{\infty} x \frac{1}{\sqrt{2\pi\sigma^2}} \exp\left(-\frac{(x-\mu)^2}{2\sigma^2}\right)$$

The normal random variable with mean zero and variance one is called a standardized normal random variable. Its probability density function is

$$\frac{e^{-x^2/2}}{\sqrt{2\pi}}. \tag{B.15}$$

Bivariate normal random variable

The bivariate normal distribution for two (correlated) random variables $\mathbf{X}$ and $\mathbf{Y}$ is characterized by two means, μ_x and μ_y, two variances, σ_x^2 and σ_y^2, and the correlation coefficient, ρ. The probability density for the bivariate normal is the daunting

$$\frac{1}{2\pi\sqrt{\sigma_x^2\sigma_y^2(1-\rho^2)}}e^{-q/2},$$

where

$$q = \frac{1}{1-\rho^2}\left[\left(\frac{x-\mu_x}{\sigma_x}\right)^2 - 2\rho\left(\frac{x-\mu_x}{\sigma_x}\right)\left(\frac{y-\mu_y}{\sigma_y}\right) + \left(\frac{y-\mu_y}{\sigma_y}\right)^2\right].$$

Fortunately, we will not have to use this formula in this book. Rather, we require only some of its moments and properties.

The expected value of $\mathbf{Y}$, given that $\mathbf{X} = x$, is called the regression of $\mathbf{Y}$ on $\mathbf{X}$ and is

$$E\{\mathbf{Y} \mid \mathbf{X} = x\} = \mu_y + \beta(x - \mu_x), \tag{B.16}$$

where the regression coefficient is

$$\beta = \frac{\mathrm{Cov}\{\mathbf{X}, \mathbf{Y}\}}{\sigma_x^2} = \frac{\rho\sigma_y}{\sigma_x}. \tag{B.17}$$

Expressed as the deviation from the mean, this becomes

$$E\{\mathbf{Y} \mid \mathbf{X} = x\} - \mu_y = \beta(x - \mu_x). \tag{B.18}$$

This is the form of the regression of y on x that is most used in quantitative genetics.

Bibliography

CAVALLI-SFORZA, L. L., AND BODMER, W. F., 1971. The Genetics of Human Populations. W. H. Freeman and Company, San Francisco.

CLAYTON, G. A., MORRIS, J. A., AND ROBERTSON, A., 1957. An experimental check on quantitative genetical theory. II. Short-term responses to selection. *J. Genetics* 55:131–151.

CLAYTON, G. A., AND ROBERTSON, A., 1955. Mutation and quantitative variation. *Amer. Natur.* 89:151–158.

DARWIN, C., 1859. On the Origin of Species by Means of Natural Selection. John Murray, London.

ENDLER, J. A., 1986. Natural Selection in the Wild. Princeton University Press, Princeton.

EWENS, W. J., 1969. Population Genetics. Methuen, London.

FALCONER, D. S., 1989. Introduction to Quantitative Genetics, 3rd ed. Longman, London.

FISHER, R. A., 1918. The correlation between relatives under the supposition of Mendelian inheritance. *Trans. Roy. Soc. Edinburgh* 52:399–433.

FISHER, R. A., 1958. The Genetical Theory of Natural Selection. Dover, New York.

GILLESPIE, J. H., 1991. The Causes of Molecular Evolution. Oxford Univ. Press, New York.

GREENBERG, R., AND CROW, J. F., 1960. A comparison of the effect of lethal and detrimental chromosomes from *Drosophila* populations. *Genetics* 45:1153–1168.

HARRIS, H., 1966. Enzyme polymorphisms in man. *Proc. Roy. Soc. Ser. B* 164:298–310.

HARTL, D. L., AND CLARK, A. G., 1989. Principles of Population Genetics. Sinauer Assoc., Inc., Sunderland.

HOULE, D., 1992. Comparing evolvability and variability of quantitative traits. *Genetics* 130:195–204.

HUDSON, R. R., 1990. Gene genealogies and the coalescent process. *Oxford Surv. Evol. Biol.* 7:1–44.

JEFFS, P. S., HOLMES, E. C., AND ASHBURNER, M., 1994. The molecular evolution of the alcohol dehydrogenase and alcohol dehydrogenase-related genes in the *Drosophila melanogaster* species subgroup. *Mol. Biol. Evol.* 11:287–304.

JOHNSON, M. S., AND BLACK, R., 1984. Pattern beneath the chaos: The effect of recruitment on genetic patchiness in an intertidal limpet. *Evolution* 38:1371–1383.

JOHNSON, M. S., AND BLACK, R., 1984. The Wahlund effect and the geographical scale of variation in the intertidal limpet *Siphonaria* sp. *Marine Biol.* 79:295–302.

KIMURA, M., 1962. On the probability of fixation of mutant genes in a population. *Genetics* 47:713–719.

KIMURA, M., 1983. The Neutral Theory of Molecular Evolution. Cambridge University Press, Cambridge.

KIMURA, M., AND OHTA, T., 1971. Protein polymorphism as a phase of molecular evolution. *Nature* 229:467–469.

KIRKPATRICK, M., AND JENKINS, C. D., 1989. Genetic segregation and the maintenance of sexual reproduction. *Nature* 339:300–301.

KONDRASHOV, A., 1988. Deleterious mutations and the evolution of sexual reproduction. *Nature* 336:435–440.

KREITMAN, M., 1983. Nucleotide polymorphism at the alcohol dehydrogenase locus of *Drosophila melanogaster*. *Nature* 304:412–417.

LANDE, R., 1976. Natural selection and random genetic drift in phenotypic evolution. *Evolution* 30:314–334.

LANDE, R., AND SCHEMSKE, D. W., 1985. The evolution of self-fertilization and inbreeding depression in plants. I. Genetic models. *Evolution* 39:24–40.

LANGLEY, C. H., VOELKER, R. A., BROWN, A. J. L., OHNISHI, S., DICKSON, B., AND MONTGOMERY, E., 1981. Null allele frequencies at allozyme loci in natural populations of *Drosophila melanogaster*. *Genetics* 99:151–156.

MAYNARD SMITH, J., 1989. Evolutionary Genetics. Oxford University Press, Oxford.

MORTON, N. E., CROW, J. F., AND MULLER, H. J., 1956. An estimate of the mutational damage in man from data on consanguineous marriages. *Proc. Natl. Acad. Sci. USA* 42:855–863.

MOUSSEAU, T. A., AND ROFF, D. A., 1987. Natural selection and the heritability of fitness components. *Heredity* 59:181–197.

MUKAI, T., 1968. The genetic structure of natural populations of *Drosophila melanogaster*. VII. Synergistic interaction of spontaneous mutant polygenes controlling viability. *Genetics* 61:749–761.

MUKAI, T., CHIGUSA, S. I., METTLER, L. E., AND CROW, J. F., 1972. Mutation rate and dominance of genes affecting viability in *Drosophila melanogaster*. *Genetics* 72:335–355.

MUKAI, T., AND COCKERHAM, C. C., 1977. Spontaneous mutation rates at enzyme loci in *Drosophila melanogaster*. *Proc. Natl. Acad. Sci. USA* 74:2514–2517.

MULLER, H. J., 1932. Some genetic aspects of sex. *Amer. Natur.* 66:118–138.

NEVO, E., BEILES, A., AND BEN-SHLOMO, R., 1984. The evolutionary significance of genetic diversity: Ecological, demographic and life history correlates. In G. S. Mani, ed., *Evolutionary Dynamics of Genetic Diversity*, 13–213. Springer-Verlag, Berlin.

ROYCHOUDHURY, A. K., AND NEI, M., 1988. Human Polymorphic Genes World Distribution. Oxford University Press, New York.

SCHEMSKE, D. W., AND LANDE, R., 1985. The evolution of self-fertilization and inbreeding depression in plants. I. Empirical observations. *Evolution* 39:41–52.

SIMMONS, M. J., AND CROW, J. F., 1977. Mutations affecting fitness in *Drosophila* populations. *Annu. Rev. Genet.* 11:49–78.

SLATKIN, M., AND BARTON, N. H., 1989. A comparison of three indirect methods for estimating average levels of gene flow. *Evolution* 43:1349–1368.

VOELKER, R. A., LANGLEY, C. H., BROWN, A. J. L., OHNISHI, S., DICKSON, B., MONTGOMERY, E., AND SMITH, S. C., 1980. Enzyme null alleles in natural populations of *Drosophila melanogaster*: Frequencies in a North Carolina population. *Proc. Natl. Acad. Sci. USA* 77:1091–1095.

WRIGHT, S., 1929. Fisher's theory of dominance. *Amer. Natur.* 63:274–279.

WRIGHT, S., 1934. Physiological and evolutionary theories of dominance. *Amer. Natur.* 68:25–53.

Index

Library of Congress Cataloging-in-Publication Data

Gillespie, John H.
Population genetics : a concise guide / John H. Gillespie
p. cm.
Includes bibliographical references and index.
ISBN 0-8018-5754-6 (alk. paper). — ISBN 0-8018-5755-4 (pbk. : alk. paper)
1. Population Genetics. I. Title.
QH455.G565 1998
576.5′8—dc21 97-19509
CIP